Progestogens in Obstetrics and Gynecology

Howard J.A. Carp
Editor

Progestogens in Obstetrics and Gynecology

Springer

Editor
Howard J.A. Carp
Department of Obstetrics & Gynecology
Sheba Medical Center
Tel Hashomer, Israel

Sackler School of Medicine
Tel Aviv University
Tel Aviv, Israel

ISBN 978-3-319-35165-0 ISBN 978-3-319-14385-9 (eBook)
DOI 10.1007/978-3-319-14385-9

Springer Cham Heidelberg New York Dordrecht London

Softcover reprint of the hardcover 1st edition 2015

Printed on acid-free paper

Springer International Publishing AG Switzerland is part of Springer Science+Business Media (www.springer.com)

Foreword

In 1930 WM Allen and GW Corner used the name progestin (later renamed progesterone in 1935) to describe a hormone which was responsible for implantation and embryo survival. They could not have imagined the profound relevance of the hormone and implications for the development of a class of drugs, which are in widespread use and affect our daily lives. For many years progesterone was thought of as a hormone only affecting pregnancy, and the wider implications of triggering the progesterone receptor were relatively ignored. In the 1960s the contraceptive pill became available and caused a revolution in the way that women could plan their pregnancies. Progestogens were introduced to modulate the estrogen used to inhibit ovulation in the contraceptive pill.

Since then a whole host of synthetic drugs known as progestogens have come into clinical use. The main use of progestogens is in pregnancy. Progestogens are used in luteal support, to prevent miscarriage, and to prevent preterm labor. Progestogens are used outside of pregnancy for abnormal uterine bleeding, cycle control, hormone replacement therapy, and even in the prevention and treatment of endometrial cancer. Today, we know that progesterone is found in nonmammalian vertebrates. Progesterone had a physiological role as an anti-inflammatory agent and neurosteroid long before mammalian pregnancy had evolved. Therefore, it is hardly surprising that progestogens are being used as possible anti-inflammatory agents in endometriosis, and even in the male in traumatic brain injury, and in multiple sclerosis.

Just as the early investigators in the 1930s could not realize the implications of their discovery, it is difficult to prophesy the future. A new field of development is receptor modulators. Mifepristone is a progesterone receptor modulator. It was introduced as an abortifacient. However, today new uses are being developed for receptor modulators. Experimental work with uterine fibroids may entirely change the management of fibroids and affect the whole approach to surgery for gynecological conditions.

This book brings together all the aspects of progestogens in gynecological (and nongynecological) practice. There are chapters governing basic scientific topics such as physiology and pharmacology. The major applications of progestogen therapy luteal support, miscarriage, preterm labor, contraception, abnormal uterine bleeding, etc. have been described in depth. However, in clinical practice, there are always controversies, leaving the clinician puzzled as how to help the patient. The different progestogens with their overlapping effects on estrogen, androgen, glucocorticoid, and mineralocorticoid receptors are described in order to allow the clinician to make the most appropriate choice of progestogen. This book will be useful for specialists including, gynecologists and endocrinologists, general practitioners, and those of associated disciplines, who wish to keep up to date and gain a comprehensive view of developments.

Andrea R. Genazzani
University of Pisa, Pisa, Italy
International Society of Gynecological Endocrinology (ISGE), Pisa, Italy
European Society of Gynecology (ESG), Prague, Czech Republic
International Academy of Human Reproduction (IAHR), Venice, Italy

Preface

The progestogens are defined as any compound which activates the progesterone receptor. This class of compounds is probably the most widely used class of drug in medical practice. Millions of women use progestogens in the contraceptive pill daily for many years. Women use progestogens widely in infertility treatment and in supporting pregnancy to prevent miscarriage. Progestogens are widely used to protect the endometrium in postmenopausal replacement therapy, cycle regulation, abnormal uterine bleeding, endometriosis, and even in endometrial carcinoma. However, the clinician is often in a quandary, as to which progestogen is most appropriate in any clinical situation. The actions of progestogens overlap with other steroid receptors. Therefore progestogens have estrogenic or antiestrogenic actions, androgenic or antiandrogenic actions, and glucocorticoid or mineralocorticoid actions. Each may have advantages or disadvantages depending on the clinical situation. However, progestogens also have side effects including stimulatory effects on the breast, possibly predisposing to breast carcinoma, breakthrough bleeding, acne mood changes, loss of libido, and dryness of the vagina.

Therefore, it was thought necessary to bring all the actions of progestogens together in one volume in order to discuss the actions and uses of progestogens in depth. This book tries to summarize some of the controversies and discuss the scientific basis for various uses of progestogens. It is hoped that this book describes the accumulating data in a way which is both scientifically sound and also clinically useful and which may improve the care of the patient.

The book is planned for general gynecologists and specialists working in the field. Each contributing author is an authority on a specific area of progestogen use. I would like to thank each author for the time and effort taken in preparing the manuscripts to make the publication of this book possible.

Tel Aviv, Israel Howard J.A. Carp

Contents

1 **Physiology of Progesterone** ... 1
Edi Vaisbuch, Offer Erez, and Roberto Romero

2 **Pharmacology of Progestogens** ... 33
Adolf E. Schindler

3 **Progestogens in Infertility Practice** ... 41
Ameet S. Patki and Alok Sharma

4 **Progestogens in Threatened Miscarriage** ... 53
Howard J.A. Carp

5 **Progestogens and Recurrent Miscarriage** ... 65
Luciano G. Nardo and Spyridon Chouliaras

6 **Progestogens in Preterm Labour** ... 79
Gian Carlo Di Renzo, Irene Giardina, Giulia Babucci,
Chiara Antonelli, Sandro Gerli, and Graziano Clerici

7 **Abnormal Uterine Bleeding** ... 93
Yochai Bar Shavit, Jaron Rabinovici, and Howard J.A. Carp

8 **Progestogens in Contraception** ... 111
Johannes Bitzer

9 **Progestogens and Endometriosis** ... 129
Howard. J.A. Carp, David Soriano, and Matityahu Zolti

10 **Progestogens and Breast Cancer** ... 149
Eitan Pe'er

11 **Progestogens in Endometrial Cancer** ... 159
Ido Laskov and Walter H. Gotlieb

12 Progestogens and the Menopause 177
Eitan Pe'er

13 Progestogens and Autoimmunity 183
Abraham Tsur, Grant C. Hughes, and Yehuda Shoenfeld

14 Progestogens in Non-gynecological Indications 191
Christa Nadjafi-Triebsch

Erratum to E1

Contributors

Chiara Antonelli Santa Maria della Misericordia University Hospital, Perugia, Italy

Giulia Babucci Santa Maria della Misericordia University Hospital, Perugia, Italy

Johannes Bitzer, M.D. Department Obstetrics and Gynecology, University Hospital Basel, Basel, Switzerland

Howard J.A. Carp, M.B.B.S., F.R.C.O.G. Department of Obstetrics and Gynecology, Sheba Medical Center, Tel Hashomer, Israel

Sackler School of Medicine, Tel Aviv University, Tel Aviv, Israel

Spyridon Chouliaras, M.D., M.R.C.O.G. Centre for Reproductive Medicine, Kenton & Lucas Wing, St Bartholomew's Hospital, London, UK

Graziano Clerici Santa Maria della Misericordia University Hospital, Perugia, Italy

Offer Erez Department of Obstetrics and Gynecology, Soroka Medical Center, Beer-Sheva, Israel

Sandro Gerli Santa Maria della Misericordia University Hospital, Perugia, Italy

Irene Giardina Santa Maria della Misericordia University Hospital, Perugia, Italy

Walter H. Gotlieb, M.D., Ph.D. Division of Gynecologic Oncology, Jewish General Hospital, McGill University, Montreal, QC, Canada

Grant C. Hughes, M.D. Division of Rheumatology, Department of Medicine, University of Washington, Seattle, WA, USA

Ido Laskov, M.D. Division of Gynecologic Oncology, Jewish General Hospital & Lady Davis Research Institute, McGill University, Montreal, QC, Canada

Christa Nadjafi-Triebsch Foederatio Medicorum Helveticorum (F.M.H.), Bern, Switzerland

Aerztekammer Nordrhein/Germany (A.E.K.N.O.), Düsseldorf, Germany

German Society of Anti-Aging Medicine e.V. (G.S.A.A.M.), Lengede, Germany

Swiss Society for Anti Aging Medicine and Prevention (S.S.A.A.M.P.), Zug, Switzerland

Luciano G. Nardo, M.D. M.R.C.O.G. Reproductive Health Group, Centre for Reproductive Health, Daresbury, Cheshire, UK

Ameet S. Patki, M.D., D.N.B., F.C.P.S., F.R.C.O.G. Fertility Associates, Mumbai, India

K. J. Somaiya Medical College and Hospital, Mumbai, India

Hinduja Surgical Hospitals, Mumbai, India

The Mumbai Obstetrics and Gynaecological Society (2014–15), Mumbai, India

Eitan Pe'er, M.D. Gyneco-Endocrinology Unit, Rambam Health Campus, Technion School of Medicine, Haifa, Israel

Jaron Rabinovici, M.D. Department of Obstetrics and Gynecology, Sheba Medical Center, Tel Hashomer, Israel

Sackler School of Medicine, Tel Aviv University, Tel Aviv, Israel

Gian Carlo Di Renzo, MD, PhD, FRCOG (hon), FACOG (hon). Santa Maria della Misericordia University Hospital, Perugia, Italy

Roberto Romero, M.D. Perinatology Research Branch, NICHD/NIH/DHHS, Bethesda, MD, USA

Perinatology Research Branch, NICHD/NIH/DHHS, Detroit, MI, USA

Department of Obstetrics and Gynecology, University of Michigan, Ann Arbor, MI, USA

Department of Epidemiology and Biostatistics, Michigan State University, East Lansing, MI, USA

Division of Intramural Research, Wayne State University/Hutzel Women's Hospital, NICHD/NIH/DHHS, Detroit, MI, USA

Adolf E. Schindler, M.D., Ph.D., M.D.(h.c.) Institute for Medical Research and Education, University Clinic Essen, Essen, Germany

Alok Sharma, M.D. Clinical Fellow Fertility Associates, Mumbai. India

Yochai Bar Shavit, M.D. Department of Obstetrics and Gynecology, Sheba Medical Center, Tel Hashomer, Israel

Yehuda Shoenfeld, M.D., F.R.C.P. Zabludowicz Center for Autoimmune Diseases, Sheba Medical Center, Tel-Hashomer, Israel

Sackler School of Medicine, Tel Aviv University, Tel Aviv, Israel

David Soriano, M.D. Tel Aviv University, Tel Aviv, Israel

Endometriosis center, Sheba Medical Center, Tel Hashomer, Israel

Abraham Tsur, M.D. Obstetrics and Gynecology, Sheba Medical Center, Tel Hshomer, Israel

Edi Vaisbuch Department of Obstetrics and Gynecology, Kaplan Medical Center, Rehovot, Israel

Matityahu Zolti, M.D. Obstetrics and Gynecology, Sheba Medical Center, Tel Hashomer, Israel

Chapter 1
Physiology of Progesterone

Edi Vaisbuch, Offer Erez, and Roberto Romero

1 Introduction

The corpus luteum was first discovered in 1672 by Reinier de Graaf and named in 1689 by Marcelo Malpighi [1]. Malpighi proposed that the corpus luteum produces the ovarian follicles and that the yellow substance, like egg yolk, serves to nourish the ovum [2].

In 1903, Fraenkel [3] demonstrated that the removal of the corpora lutea of rabbits before implantation, prevented implantation. Moreover, luteectomy in early pregnancy (<14 days) resulted in pregnancy loss [3]. In 1929, Corner and W.M. Allen [4] described that injecting extracts of the corpus luteum into castrated adult female rabbits induces a characteristic alteration of the endometrium identical to progestational proliferation, previously shown to be due to the presence of corpora lutea in the ovaries [4]. In a second report, Allen and Corner demonstrated

E. Vaisbuch
Department of Obstetrics and Gynecology, Kaplan Medical Center, Rehovot, Israel

O. Erez
Department of Obstetrics and Gynecology, Soroka Medical Center, Beer-Sheva, Israel

R. Romero, M.D. (✉)
Perinatology Research Branch, NICHD/NIH/DHHS, Bethesda, MD, USA

Perinatology Research Branch, NICHD/NIH/DHHS, Detroit, MI, USA

Department of Obstetrics and Gynecology, University of Michigan, Ann Arbor, MI, USA

Department of Epidemiology and Biostatistics, Michigan State University, East Lansing, MI, USA

Division of Intramural Research, Wayne State University/Hutzel Women's Hospital, NICHD/NIH/DHHS, 3990 John R., Box #4, Detroit, MI 48201, USA
e-mail: romeror@mail.nih.gov

H.J.A. Carp (ed.), *Progestogens in Obstetrics and Gynecology*,
DOI 10.1007/978-3-319-14385-9_1

that in ovariectomized rabbits (at the 18th hour of pregnancy), the presence of progestational proliferation induced by corpus luteum extracts may sustain normal implantation as well as embryo survival and growth; whereas in the absence of progestational proliferation the embryos never survived beyond the fourth day [5]. Therefore, the extracts of corpus luteum were essential for both implantation and early pregnancy maintenance.

In 1930, Allen proposed the name "progestin" to refer to the hormone responsible for these biological effects [6], and the race for the isolation of this hormone ended in 1934 when four different groups reported its purification and characterization [7–10]. Each group suggested a different name to refer to the main corpus luteum hormone, and the name "progesterone" came by consensus after a meeting of the League of Nation's Health Organization in 1935 [11].

The major target organ of progesterone is the reproductive tract, however, progesterone has a systemic effect and influences other organs including, but not limited to: the mammary glands, the nervous system and the brain, the heart, the bone, the endocrine and the immune systems (Table 1.1) [12–16]. In the reproductive system, progesterone, in association with estrogen, is involved in the development and sexual maturation of the reproductive organs and orchestrates the menstrual cycle [12, 17–19]. This chapter describes the specific effects of progesterone on the uterus (myometrium and endometrium) and the cervix during normal menstrual cycle and pregnancy. Detailed discussion on the effect of progesterone on organs outside the female reproductive tract is described in other chapters.

Table 1.1 Tissue and cell types expressing progesterone receptors

Tissue	Cell type
Uterus	Endometrium
	Myometrium
Ovary	Luteinizing granulose
	Preovulatory granulose
	Corpus luteum
Other reproductive organs	Vagina
	Testes
Breast	Ductal and lobular epithelial cells
Brain	Pituitary
	Hypothalamus
	Preoptic area
Others	Vascular endothelium
	Thymus
	Pancreatic islets
	Osteoblast-like cells
	Lung

Modified from Graham JD, Clarke CL. Endocrine Reviews 1997;18502–519

2 The Mechanisms of the Cellular Action of Progesterone

Progesterone can evoke genomic or non-genomic responses upon its interaction with target cells. The term "*genomic actions*" refers to the cellular response involved in the activation of the genetic machinery, resulting in modulation of DNA expression. The genomic actions of progesterone are largely, but not only, mediated by the progesterone receptor (PR) [20]. The term "*non-genomic actions*" indicates the cellular responses to progesterone which involve alternative pathways, such as the activation of signal-transduction cascades, the generation of intracellular second messengers, and the modulation of protein kinases and ion fluxes (Fig. 1.1) [21, 22].

2.1 Genomic Actions of Progesterone and the Cytosolic Progesterone Receptor

The classical cytosolic PR [20], a member of the steroid/nuclear receptor superfamily, is a mediator of the genomic actions of progesterone. In resting conditions, this receptor is localized in the cytosol, within a large complex of proteins, including heat shock proteins and FK506-binding proteins, contributing to maintaining it in a transcriptionally inactive state (Fig. 1.2) [23].

This cytosolic receptor can be activated by ligand-dependent [23], and ligand-independent mechanisms [24]. In the ligand-dependent pathway, progesterone

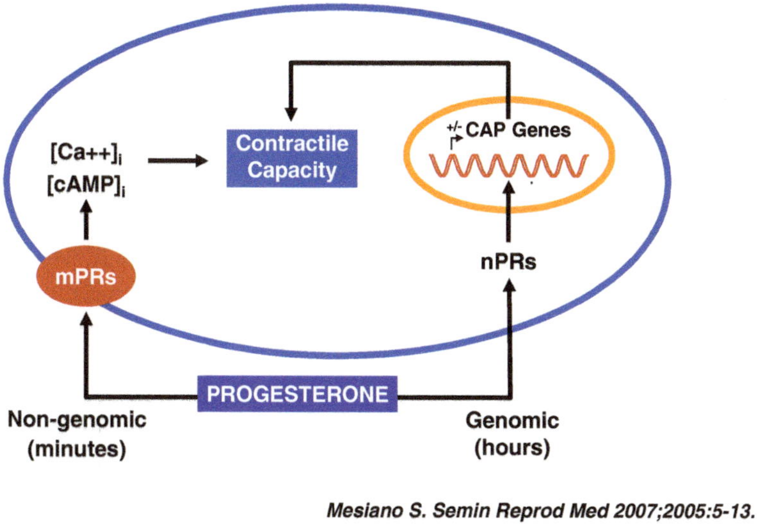

Fig. 1.1 Central paradigm for genomic and nongenomic progesterone actions on myometrial cells (Reproduced with permission from Thieme Publishers: Mesiano, S., *Myometrial progesterone responsiveness*. Semin. Reprod. Med., 2007. **25**(1): p. 5–13 [149])

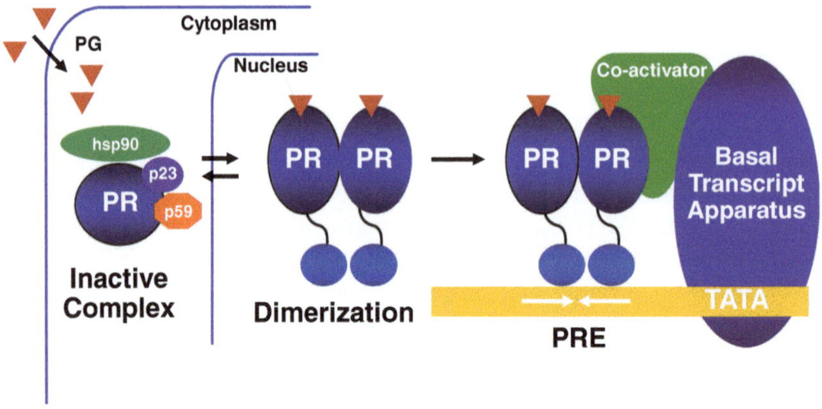

Fig. 1.2 Progesterone activation of the cytosolic progesterone receptor (Reproduced with permission from Elsevier: Leonhardt, S.A., Boonyaratanakornkit, V., Edwards, D.P. *Progesterone receptor transcription and non-transcription signaling mechanisms*. Steroids, 2003. **68**(10–13): p. 761–770 [234])

gains access into the cell through passive diffusion or facilitated transport, and binds to the receptor, which changes its conformation (including dimerization and shedding of the heat shock proteins) [25]. This process allows the dissociation of the PR from the chaperone complex, its translocation into the nucleus, and finally its interaction with the DNA, where binding in a homodimeric form to cis-acting DNA progesterone response elements (PREs) modulates the transcription of target genes [23]. The ligand-independent activation of the PR is, instead, the result of cross-talk between membrane receptors and intra-cellular kinases, including a cAMP-dependent kinase, the cyclin A/cyclin-dependent kinase-2 (Cdk2), the mitogen activating protein kinase (MAPK), the stress-activated p38 MAPK, and the protein kinases A and C [26].

The gene encoding the human PR is located on chromosome 11q22.1 and has eight exons. Alternative splicing allows for the synthesis of different isoforms of the receptor [27]. The two major isoforms of the PR are progesterone receptor-A (PR-A) and B (PR-B). These isoforms, although characterized by a different length, do not differ in their amino acid sequence: PR-B is 933 amino acids in length, while PR-A lacks 164 amino acids at the amino terminus. *In vitro*, PR-B is a stronger trans-activator than PR-A, whereas PR-A acts as a trans-repressor of PR-B and of other steroid receptors [28, 29]. Structurally, both isoforms consist of an amino-terminal region, a centrally-located DNA binding domain, and a carboxy-terminal hinge region containing nuclear localization signals, as well as the ligand-binding domain. Three transcription activation function (AF) domains have been identified within the PR amino acid chain. AF-1 is located upstream of the DNA binding domain, while AF-2 is located in the ligand-binding domain [30]. AF-3 is unique to the PR-B isoform, and is located within the N-terminal region [28, 31]. In addition, an inhibitory function region, located between the AF-1 and AF-3 domains, has been proposed to be responsible for the auto-inhibition and trans-repression of the PR [32]. Interestingly, most of the evolutionary changes in the human PR took place in this region [33]. (For more information on the structure of the Human Progesterone Receptor gene go to http://www.ncbi.nlm.nih.gov/gene/5241).

A third isoform of the PR, the PR-C, was also described [34–36]. This is a 60 kDa N-terminally truncated isoform, lacking the DNA binding domain, but containing the hormone binding region with the sequence for dimerization and nuclear localization [34, 35, 37]. The cytoplasmatic PR-C has been suggested to inhibit PR-B activity by sequestrating the locally available progesterone [37]. The nuclear PR-C can form heterodimers with PR-B, therefore interfering with its binding to the response elements in the DNA [37]. In contrast, PR-C can enhance the progestin-induced transcriptional activity of PR-A and PR-B isoforms, either by sequestrating the co-repressors and/or by increasing the capacity of the heterodimers of PR-A or PR-B with PR-C to recruit co-activators [35]. In this manner, PR-C could be involved in the modulation of the transcriptional activity of PR-A and PR-B, contributing to the pleiotropic effects of progestins [35]. Additional isoforms of the PR, such as PR-S [38] and PR-M [39] have also been identified and partially characterized. It has been proposed that the tissue responses to progesterone may be affected by changes in the expression ratio of the different isoforms [36].

Importantly, the validity of the immunoassay that was used in the identification of some of the PR isoforms, such as PR-C and PR-M, has been questioned [40, 41]. It has been demonstrated that the nuclear PR antibodies used may cross-react with cytoskeletal proteins (α-actinin, desmin and vimentin), suggesting that these antibodies are not specific for these PR isoforms [41].

2.2 The Role of Co-regulators in Progesterone Signaling

The activity of the nuclear PR is regulated not only by the hormone itself but also by co-regulators (co-activators and co-repressors) as well as by chromatin modifiers [42]. Co-regulators can enhance or inhibit gene transcription by creating a functional link between the ligand-activated receptors, the DNA and the transcription factors [43]. The existence of "intermediary factors" in the PR nuclear signaling was described more than four decades ago by the group of O'Malley [44]. Since then, the interest in co-regulators has increased, because of their possible involvement in the "transcriptional interference" in the tissue-specific responses evoked by nuclear receptors ligands, selective receptor modulators (i.e., Tamoxifen and Raloxifene), and their role in the pathogenesis/progression of neoplastic disease [45]. Thus, the possible involvement of progesterone co-regulators in the modulation of myometrial progesterone action should be taken into account [46, 47].

Progesterone co-activators include members of the "Steroid Receptor Co-activator" (SRC/p160) family [48], such as SRC-1, SRC-2, and SRC-3, which share a strong sequence homology [48, 49]. An involvement of these progesterone co-activators in normal growth, puberty, and female reproductive function, as well as in mammary gland development, is supported by studies on genetically-modified animals. SRC-1 [50, 51] is an important co-activator in the uterus, whereas SRC-3 is in the mammary gland [51, 52], and SRC-2 in both organs [46, 47]. Of note, SRC-1 and SRC-2 knockout mice manifested a deficient uterine response to progesterone stimulation. However, SRC-1 knockout mice preserved their fertility [50], whereas SRC-2 knockout mice had an early block of embryo implantation [46]. Progesterone receptor co-activators share a NRbox (also called LXXLL motif) that is necessary for binding to the "co-activator binding groove" in the receptor [49, 53].

Co-repressors of the progesterone receptors inhibit transcription factors recruitment and down-regulate the receptor-dependent gene expression. This is accomplished preferentially by recruiting histone deacetylases [54], which enhance tight nucleosome-DNA interactions and increase chromatin compaction [49]. However, the molecular basis of the interactions between steroid receptors and co-repressors is not well-defined [49].

2.3 *Non-genomic Actions of Progesterone*

The identification of steroid receptors on cells lacking a functional nucleus (i.e., spermatozoa, erythrocytes and platelets) supports the hypothesis of the existence of non-genomic steroid actions. This is a fast-track response system, and in contrast to the long response time (i.e., hours/days) of the "genomic" pathways, the cellular responses evoked by the "non-genomic" pathways are rapid [21, 22]. The first evidence in support of the existence of non-genomic progesterone actions came from the study of progesterone responses in germ cells such as oocytes and spermatocytes. Some of the non-genomic actions exerted by progesterone on these cells include changes in intracellular calcium concentrations [55–58], promotion of Na^+ [59] and Cl^- [60] fluxes, inhibition of adenylate cyclase activity with a consequent decrease in intracellular cAMP levels [61], and the involvement in G proteins-phospholipase C-inositol trisphosphate, and diacylglycerol signaling [62, 63].

"Membrane-initiated steroid signaling" defines the non-genomic activities of progesterone that are secondary to activation of membrane-localized progesterone receptors [21, 22]. Evidence that at least some of the non-genomic actions of progesterone are mediated by membrane located receptors include: 1) progesterone application outside the cell is more effective in decreasing intracellular cAMP concentrations than upon its cytoplasmic microinjection [64]; 2) progesterone activity is sustained after conjugation with synthetic polymers [65, 66] or its covalent binding to large molecules, such as albumin [62, 67], which prevent progesterone access into the cytosol; 3) progesterone effects are reduced in the presence of antibodies directed toward progesterone membrane-binding proteins [58]; and 4) the non-genomic activities of progesterone, such as Ca^{2+} influx, are not affected by inhibitors of genomic progesterone responses, including RU38486 and RU486 [55–57].

Progesterone high-affinity binding proteins and receptors have been identified on the cellular membranes of a variety of cells such as spermatozoa [55, 57, 58], porcine liver microsomes [68], porcine vascular muscle cells [69], as well as the brains of female mice knocked out for the classical PR [70]. Some researchers have previously argued that the existence of a progesterone binding site does not necessarily indicate that the receptor is functionally active in terms of cellular signaling, and that a characterization of non-classical receptors is still required [21, 22].

Non-genomic progesterone receptors display different affinities, binding capacities, and dose response/competition curves for progesterone and other molecules sharing progestin structure. For example, the recombinant human mPRγ, produced in an E. coli expression system, has a high affinity, saturable, single binding site for progesterone and several of its hydroxylated derivatives; however, recombinant human mPRγ does not bind and has no affinity for synthetic progestins and anti-progestins [71]. Similarly, there is evidence indicating the presence of at least two distinct membrane surface progesterone receptors in capacitated human spermatozoa: a high affinity site that is specific for progesterone, and a low affinity site that binds with equal affinity to 11β-hydroxyprogesterone and 17α-hydroxyprogesterone [57].

3 The Physiologic Effects of Progesterone

3.1 The Effect of Progesterone on the Immune System

The effect of progesterone on the immune system deserves special attention. The immune system in the female reproductive tract faces two opposing challenges: the consistent exposure to infectious pathogens, and in contrast, the need to be tolerant to both the allogenic spermatozoa and the semi-allogenic fetus. To overcome these challenges, the female sex steroids (i.e., estrogen and progesterone) control the function of the innate and adaptive immune systems in the reproductive tract according to the changes along the menstrual cycle and during pregnancy [72–74]. Indeed, in the rat uterus, major histocompatibility complex (MHC) class II positive cells, macrophages, granulocytes and dendritic cells were more abundant in the endometrial stroma and around uterine glandular epithelium in the estrus stages of the menstrual cycle relative to the diestrus stages in which progesterone is the dominant hormone [75]. Moreover, ovariectomy in mice results in a decrease in the number of uterine macrophages that can be restored by hormonal treatment [76].

The uterine/decidual natural killer (NK) cells are affected by progesterone as well: these cells have a role in promoting blastocyst implantation and maintenance of pregnancy [77, 78]. During the mid-late luteal phase the numbers of this unique population of NK cells is elevated [79, 80] as a result of the increased decidual concentrations of interleukin (IL)-15, and IL-15 mRNA [81]. Their number further increases during early stages of pregnancy and decrease from mid-gestation to term [79].

The immunologic recognition of pregnancy also leads to a higher expression of PR on membrane of uterine NK cells [82] and decreased cytotoxic activity in comparison to the non-pregnant state [83].

Progesterone diverts the T cell response toward Th-2 rather than Th-1, leading to higher secretion of IL-6 and IL-10, as well as supporting B-cell antibody production [84–86]. During pregnancy there is also a change in the antibody population and a shift toward non-symmetric "blocking" antibodies (those glycosilated by mannose-rich oligosaccharide only on one of the Fab regions and although can combine with antigen, they poorly activate phagocytosis, complement fixation, and cytotoxicity) and their prevalence increases from 9 % in the non-pregnant state to 29 % in pregnant women [87]. Yet, they can compete with symmetric and competent antibodies that have the same specificity and prevent their action. This was suggested as a mechanism to reduce the antibodies' mediated response against the invading trophoblast during pregnancy and to control the equilibrium of maternal anti-fetal immune responses [88].

It has been reported that the effects of progesterone on the T cell response, B cell activity, generation of non-symmetric antibodies, and NK cytotoxicity is mediated by progesterone induced blocking factor (PIBF) (Fig. 1.3), a 34 kDa immunoregulatory

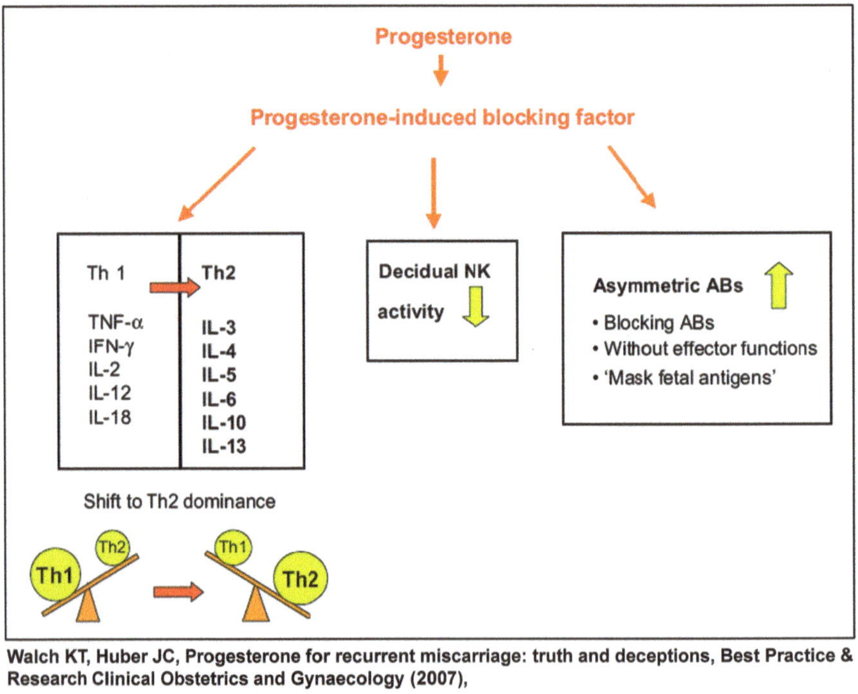

Fig. 1.3 The effect of progesterone and progesterone induced blocking factor (PIBF) on maternal immune system during pregnancy (Reproduced with permission from Elsevier: Walch, K.T. and J.C. Huber, *Progesterone for recurrent miscarriage: truth and deceptions*. Best Pract Res Clin Obstet Gynaecol, 2008. **22**(2): p. 375–389 [75])

protein synthesized by PR positive lymphocytes and CD56+ decidual cells [89]. The actions of PIBF include: 1) enhancement of the production of asymmetric antibodies [90]; 2) bias the T helper response toward Th-2 activity, resulting in increased concentrations of IL-3, IL-4 and IL-10 as well as decreased IL-12 production [91]; 3) the latter combined with the inhibition of perforin secretion by PIBF in a dose-dependent manner reduces the cytotoxic activity of NK cells [91, 92]. In summary, progesterone affects all arms of the immune system and propagates the maternal tolerance to the semi-allogeneic fetus.

3.2 The Role of Progesterone in Non-pregnant Women

3.2.1 Progesterone and the Menstrual Cycle

Progesterone participates in control of ovulation, preparation and stabilization of the endometrium before implantation, regulation of the implantation process and maintenance of pregnancy [93]. During the follicular phase of the menstrual cycle, estrogen predominates and has a major role in the proliferation of the endometrium while progesterone concentrations are relatively low. Progesterone predominates during the secretory phase (maximal concentrations in the mid-luteal phase), inhibits the endometrial proliferation induced by estrogen, and changes the endometrial morphology to the secretory type [94]. However, the glandular and vascular elements continue to grow, resulting in progressive tortuosity [94]. Progesterone stimulates glycogen vacuole formation within glandular cells, resulting in the active secretion of glycoproteins and peptides by the glands into the endometrial cavity, as well as edema of the endometrial stromal tissue [95]. In the mid-luteal phase progesterone is responsible for the transformation of stromal cells into decidual cells, which is critical for the establishment of pregnancy. In the absence of pregnancy, the demise of the corpus luteum exerts a physiological progesterone withdrawal resulting in menstruation [93].

Previous exposure to estrogen is essential to stimulate synthesis of PR in endometrial cells. This enables progesterone to exert its anti-estrogenic effect on the endometrium [96] through several proposed potential mechanisms, such as down regulation of estrogen receptor expression [97], conversion of estradiol to a less active form (estrone sulphate) via the stimulation of 17-hydroxysteroid dehydrogenase and sulfotransferase [98], and suppression of estrogen-mediated synthesis/secretion of specific proteins (e.g., transcription of the proto-oncogene c-fos mRNA) [99].

In addition, progesterone increases the expression of tissue factor (TF) and plasminogen activator inhibitor-1 during decidualization [100]. It has been suggested that an increase in decidual TF concentrations is needed to secure rapid hemostasis during blastocyst implantation and placentation as well as the controlling of postpartum hemorrhage [100]. The association between the decidual expression of TF and progesterone was established by the differences in TF expression in confluent stromal cell cultures derived from proliferative phase endometrium. Stromal cell cultures that were treated with mifepristone did not increase their TF expression; moreover, administration of mifepristone to cell cultures previously exposed to estradiol + MPA or estradiol + progesterone decreased their TF content and TF mRNA expression [101]. Therefore, a low progesterone concentration could contribute to less effective decidual hemostasis, which may lead to increased decidual bleeding and a subsequent spontaneous abortion or preterm delivery.

3.2.2 Progesterone and the Myometrium in the Non-pregnant Uterus

Uterine contractile activity throughout the menstrual cycle is partially regulated by estrogen and progesterone [102–104]. This has been proposed to be mediated by cyclic changes in estrogen and progesterone receptor expression in the endometrium and sub-endometrium [105]. The decrease in the progesterone concentration in the transition from the luteal phase of one menstrual cycle to the follicular phase of the subsequent cycle is followed by increased uterine contractility, which aids in clearing menstrual contents [104]. The rise in estrogen concentration during the late follicular phase further increases uterine contractility, preparing the uterus to facilitate sperm motility toward the fallopian tube [104]. During the luteal phase, following an increase in progesterone concentration, the uterus is relatively quiescent [102, 104].

Of note, studies in non-pregnant women demonstrated that plasma progesterone concentrations do not reflect the actual progesterone concentrations in the myometrium. Akerlund et al. [106] measured the estrogen and progesterone concentrations in non-pregnant uteri of women with normal menstrual cycles, and demonstrated that there is no correlation between plasma and tissue progesterone concentration in the same individual, although the progesterone concentrations in the plasma and myometrial tissue change during the menstrual cycle. The authors reported that in peri- and postmenopausal women, the myometrial concentration of progesterone remains comparable to those in menstruating women, despite a substantial decline in plasma concentrations of this hormone, suggesting that the myometrial uptake of ovarian hormones may be saturated even if plasma concentrations are relatively low. Moreover, the myometrial/plasma ratio of progesterone decreases significantly during the luteal phase [106]. This may be due to down-regulation of the myometrial progesterone receptors following accumulation of progesterone in the myometrium [107].

3.2.3 The Effect of Progesterone on the Uterine Cervix During the Menstrual Cycle

The uterine cervix is a primary end organ that is responsive to pubertal hormonal action [108], cyclical changes in sex hormones during the menstrual cycle, pregnancy, labor and menopause [109–111]. The expression of PR changes significantly in the glandular epithelium of the cervix, reaching its peak in the early secretory phase and declining sharply afterwards [112]. Progesterone has a dramatic effect on the constituents of the cervical mucus, and hence, on the function of cervical secretions [113, 114]. The cervical mucus was defined by Odeblad et al. [113–115] as type E (estrogenic) and type G (progestogenic), the latter a thick and sticky mucus. Indeed, one of the suggested mechanisms by which progestins exert their contraceptive effect is through changes in the chemical properties of mucus [116, 117].

The available data suggest that progesterone has an effect on the cervix in the non-pregnant state; however, how and to what extent this effect is important in physiologic and pathologic conditions has yet to be determined.

3.3 *Progesterone and Pregnancy*

3.3.1 The Role of Progesterone in the Maintenance of Normal Pregnancy and Parturition

Estrogen and progesterone play a central role in pregnancy [118]. The corpus luteum is the main source of progesterone until the seventh week of gestation, then the placenta takes over as the main source of progesterone between 7 and 9 weeks of gestation, a transition termed the "luteal-placental shift" [119, 120] (Fig. 1.4). Indeed, ovariectomy before 8 weeks of gestation results in abortion, but has no effect on the pregnancy if performed after 9 weeks of gestation [121]. Maternal

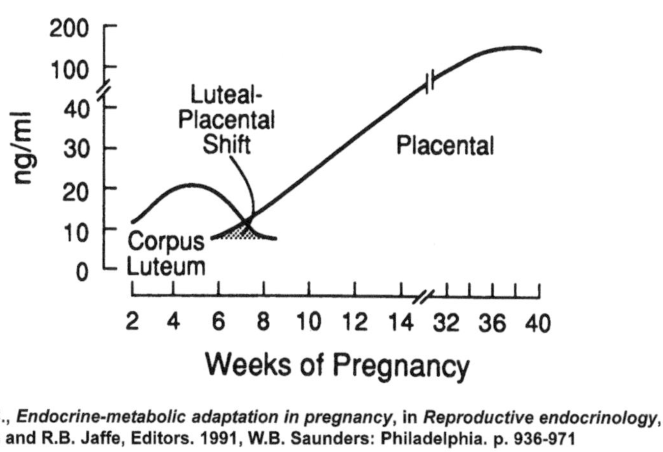

Fig. 1.4 The luteal-placental shift in progesterone production during pregnancy (Reproduced with permission from Elsevier Ltd.: Yen, S.S.C., *Endocrine-metabolic adaptation in pregnancy*, in *Reproductive endocrinology*, S.S.C. Yen and R.B. Jaffe, Editors. 1991, W.B. Saunders: Philadelphia. p. 936–971[235])

plasma progesterone concentrations rise during pregnancy from 40 ng/ml in the first trimester to 160 ng/ml in the third trimester [120]. At term, the placenta produces approximately 250 mg of progesterone per day, of which 90 % is secreted to the maternal circulation and only 10 % into the fetal circulation. However, the fetal plasma progesterone concentration is sevenfold higher than the maternal, probably due to the differences in their volume of distribution [122]. There are contradictory reports concerning the changes in amniotic fluid progesterone concentration during pregnancy: while one group reported that progesterone increases during gestation [123], others found that amniotic fluid progesterone concentration is higher in the second trimester than at term [124]. Of interest, the process of parturition at term [125] or preterm [126] was not associated with significant changes in amniotic fluid progesterone concentrations.

During pregnancy, progesterone is thought to maintain myometrial quiescence and inhibit cervical ripening, while estrogens have been implicated in increasing myometrial contractility and excitability, as well as in the induction of cervical ripening prior to the onset of labor [127–129]. However, before spontaneous parturition, the changes in sex steroid serum concentrations differ between different species. In many species, a fall in maternal serum progesterone concentration occurs prior to the onset of parturition, but the mechanism for this "progesterone withdrawal" depends, to a large extent, on whether or not the placenta or the corpus luteum is the main source of progesterone [130].

Luteolysis is a crucial component in the mechanism of parturition in the rat, mouse and rabbit [131–133]. An increase in local progesterone metabolism in both the uterus [134] and cervix [135] was associated with the onset of labor in mice. In sheep and goats, an increase in fetal plasma cortisol induces the placental production of P450 C17 enzymes (17α-hydroxylase and C17–20 lyases), which catalyze the conversion of progesterone to androstenedione, which is transformed into estrogen by aromatases [130]. In goats, the corpus luteum is responsible for most maternal progesterone production, and luteal regression is still required before parturition [133]. However, in primates (including humans) and guinea pigs there is no apparent change in circulating maternal progesterone concentration before parturition. The human placenta, in contrast to that of sheep, lacks P450 C17 enzymes, and therefore, cannot synthesize estrogen and androstenedione from C21-progestins, thus, progesterone is the final product of the human placenta.

A serum "progesterone withdrawal" was not demonstrated in humans or guinea pigs; yet, progesterone is considered important in pregnancy maintenance because inhibition of its action could result in parturition in both species. Administration of anti-progestins [i.e., mifepristone or onapristone] to pregnant women [136], primates [137] or guinea pigs [128] can induce abortion and/or labor [118]. Alternative mechanisms for the suspension of progesterone action without a serum progesterone withdrawal have been proposed, including: 1) binding of progesterone to a high affinity protein that reduces the functional active form [138]; 2) an increase in cortisol concentration during late pregnancy which may compete with progesterone binding to the glucocorticoid receptors, resulting in a functional progesterone withdrawal [139]; and 3) the conversion of progesterone to an inactive form within the target cell before interacting with its receptor. Indeed, human amnion and chorion can convert progesterone to the inactive 20α-dihydroxyprogesterone, and this metabolite increases with gestational age and around the time of parturition [140, 141]. However, none of these hypotheses have been proven [142]; therefore, the focus of investigation has shifted to the abundance and modulation of estrogen-progesterone receptor expression, and progesterone binding capability to its nuclear response element.

3.3.2 The Nuclear Progesterone Receptor in the Myometrium During Pregnancy and Parturition

Conflicting results have been reported regarding the role of the PR in the myometrium during human pregnancy and parturition [142–145]. These discrepancies have been partially attributed to the existence of multiple receptor isoforms, whose myometrial expression is spatially and temporally regulated throughout gestation [146]. Thus, it is likely that the results of the studies may be affected by the sampling site and the specificity of the assay. Of note, initial studies on PR expression in the human myometrium did not distinguish between the different isoforms and were performed on biopsies isolated from the lower uterine segment. In contrast, more recent studies tested the expression of the different receptor isoforms and focused preferentially on the fundal myometrium. The latter is more likely to reflect the molecular changes that mediate uterine contractility than the lower uterine segment, which reacts in favor of dilatation [36]. Finally, the non-genomic progesterone actions have broadened the research on the mechanism of labor toward identification of membrane progesterone receptors that may participate in the suspension of progesterone action.

The key mechanisms explaining the functional progesterone withdrawal include either a reduction of the total number of progesterone receptors within the target tissue, or a relative increase of inhibitory PR isoforms. Rezapour et al. [143] investigated the expression of progesterone receptors in the myometrium of women at term not in labor and in the active phase of spontaneous labor and found significant changes in the distribution of their receptors after the onset of labor. The active normal labor group had a higher receptor concentration in the upper uterine segment, as well as a higher upper to lower uterine segment receptor ratio than the not-in-labor group. Of interest, myometrial PR concentrations were lower in oxytocin-resistant labor than in normal labor. The authors suggested that although progesterone is involved in labor-associated changes in the myometrium through receptor mediated processes, it is not an inhibitor of myometrial contractility and, thus, not consistent with the progesterone withdrawal theory; however, this study did not discriminate between the different progesterone receptor isoforms. Supporting these findings are the *in vitro* reports that progesterone stimulates myometrial tonus and frequency of contractions [147], and has an anti-tachyphylactic effect on oxytocin-induced myometrial contractions [148].

In contrast, How et al. [144] supported the concept that a decrease in myometrial progesterone receptors may play a role in the onset of term and preterm parturition. Using myometrial samples from the lower uterine segment of women undergoing cesarean deliveries at term and preterm (in labor and not in labor), the authors found that in the pregnant myometrium, PR-A was relatively more abundant than PR-B. The intensity of immunostaining for progesterone receptors was lower in term than in preterm women not in labor and in preterm and term patients in labor than those not in labor. These local changes in myometrial PR expression with advancing gestation, and with labor, have been interpreted as the mechanism through which a functional progesterone withdrawal may occur, despite the absence of changes in the peripheral or myometrial concentrations of progesterone [144].

Pieber et al. [142] analyzed the labor-associated changes in the expression of PR-A and PR-B in myometrial samples obtained during term cesarean sections from women not in labor and in labor. While PR-A expression was detected only in the presence of effective labor, PR-B was equally expressed in labor and not in labor samples. Transient transfection of myometrial cells with PR-A and PR-B confirmed that the over-expression of PR-A has a dominant repressive effect on transcription of progesterone sensitive genes within human term myometrial cells. The authors interpreted that the expression of PR-B occurs throughout gestation and is required for pregnancy maintenance, whereas a higher expression of PR-A in the presence of effective labor at term may contribute to "functional progesterone withdrawal" [142]. However, in commenting on this study, Mesiano [149] noted that the immunoblotting data reported by Pieber et al. [142] demonstrated that the abundance of PR-B exceeds that of PR-A in laboring myometrium and that most of the *in vitro* studies showed that PR-A repression of PR-B transactivation occurs only when the PR-A/PR-B ratio is greater than one. Similar results were reported by Merlino et al. [150], which determined the PR-A and PR-B concentrations and their cellular localization in the lower uterine myometrium from preterm (not in labor) and term (not in labor and in labor) cesarean deliveries. The PR-A/PR-B ratio was significantly higher in term versus preterm myometrium; moreover, the ratio further increased in labor at term. The increase in the expression of the inhibitory PR-A [142, 151], and in the PR-A/PR-B ratio in human myometrium [150], was interpreted as the possible underlying mechanism of the "functional progesterone withdrawal".

Mesiano et al. [152] demonstrated that the change in the PR-A/PR-B ratio occurs at the mRNA level. The authors compared the abundance of mRNAs encoding for PR-A and PR-B, as well as estrogen receptors (ER) α and β, in the lower uterine segment myometrium of women at term in labor and not in labor. The mRNA levels of ERα and of the homeobox gene HOXA10 were used as markers of progesterone responsiveness since their myometrial expression had been previously shown to be inhibited by progesterone. In laboring myometrium, the mean relative abundance of mRNAs encoding for PR-A, PR-B, and ERα was significantly increased than in non-laboring tissue, whereas ERβ was low and did not differ between the groups. There was a significant two to threefold increase in the PR-A/PR-B ratio in laboring compared with non-laboring specimens. Of interest, in non-laboring myometrium, the PR-A mRNA levels and the PR-A/PR-B mRNA ratio positively correlated with mRNA of ERα and HOXA10 in laboring myometrium. These positive correlations were interpreted as an indicator that progesterone responsiveness is inversely related to the PR-A/PR-B gene expression ratio and decreases at the onset of labor. Moreover, ERα could be an early gene, whereas HOXA10 a late gene responding respectively to changes in the PR-A/PR-B expression ratio. The positive correlation detected in non-laboring myometrium between ERα mRNA levels and those of contraction-associated genes, such as cyclooxygenase-2 (COX-2), and the oxytocin receptor, suggests that the process of human parturition is initiated within myometrial cells well before the onset of active labor [152].

Changes in the ratio of the PR isoforms during labor have also been reported in non-human primates. Haluska et al. [153] quantified the PR isoforms and PR mRNA expression in myometrium, decidua, and fetal membranes from rhesus monkeys during pregnancy and in spontaneous labor at term. No changes in the total PR expression were detected in the myometrium during pregnancy and labor, but there was a significant shift from PR-B dominance at mid-pregnancy to PR-A dominance in labor. In addition, both PR-A and PR-B isoforms and PR nuclear staining were nearly undetectable in amnion obtained during labor. The authors concluded that a shift to PR-A dominance in myometrium at term, together with a loss of PR in fetal membranes, may be the mechanisms for a "functional progesterone withdrawal" [153].

A possible role for prostaglandins (PGs) in the changes of PR isoforms during parturition has been suggested. Madsen et al. [154] proposed that the exposure to PGE_2 and $PGF_2\alpha$, acting via the protein kinase C pathway, may facilitate functional progesterone withdrawal by increasing the myometrial PR-A/PR-B expression ratio. In addition, the membrane progesterone receptors (mPRs) in the human myometrium may also play a role in the mechanism of a functional progesterone withdrawal during labor [155].

3.3.3 The Membrane Progesterone Receptor During Pregnancy and Parturition

The first report on the existence of high affinity membrane-associated progesterone binding sites dates back to the study of Haukkamaa et al. [156] in 1984, on microsomes prepared from pregnant and non-pregnant uteri. Although at that time, the physiological function of these receptors was unknown, the study provided evidence that mPRs exist within the uterine tissue. It was noted that such receptors differ from their soluble cytosolic counterparts, previously identified in the human uterus, in terms of the specificity of their ligands.

Labor and sex-steroids differentially modulate the mPRs. Karteris et al. [155] reported the expression of two different functional mPRs (mPRα and mPRβ) in the myometrial cells of pregnant humans that are directly coupled to G-inhibitory proteins. This results in the inhibition of adenyl cyclase, a subsequent decline in cAMP concentrations and increased phosphorylation of the myosin light chain, which facilitates myometrial contractions. The authors proposed that, during labor, progesterone acts preferentially on its membrane receptors, a modus operandi that promotes the shift from quiescence to a contractile state. This change results from the altered PR-B/PR-A ratio, the changes in sex-steroids, and the existence of complex cross-talk between the nuclear and membrane progesterone receptors [155].

Fernandes et al. [157] combined bioinformatic analyses with the expression profile of mPRs to define their role in cycling human endometrium and gestational tissues. Sequence analysis suggested that these receptors belong to the "progestin and adiponectin receptors" family. The onset of parturition was associated with a marked reduction in myometrial mPRα and mPRβ transcripts. Of interest, the levels of mPRα expression were high in the placenta, and inversely correlated with that of the nuclear PR, indicating that mPRα may have an important functional role, particularly in reproductive tissues expressing low levels of nuclear PR [157].

3.3.4 Progesterone Oxytocin Responsiveness and Ca^{2+} Fluxes

Progesterone reduces the myometrial responsiveness to oxytocin through genomic [158] and non-genomic [159, 160] pathways. In 1929, Knaus [161] reported that "when an extract of the corpus luteum is injected into rabbits, the uteri of these animals subsequently suspended in physiological solution, fails to react to the oxytocic hormone of the posterior pituitary lobe". This suggests that the corpus luteum extracts, described later as containing progesterone, induce uterine refractoriness for the uterotonic effect of oxytocin. After nearly nine decades, the exact mechanisms by which progesterone blunt uterine responsiveness to oxytocin is still not clear, and three potential mechanisms have been proposed: 1) progesterone represses the oxytocin receptor synthesis through its genomic action [158]; 2) direct interaction between progesterone and its metabolites with the oxytocin receptor [162, 163]; or 3) the continuous presence of intracellular high progesterone concentrations may alter the responsiveness of the oxytocin receptor through non-genomic effects [164].

The oxytocin receptor needs a cholesterol-rich microenvironment to become stable in its high-affinity state [165]. The intracellular binding of progesterone to the multi-drug-resistance P-glycoprotein interfere with cholesterol transport to and from the plasma membrane, and higher intracellular concentrations of progesterone inhibit cholesterol esterification as well [166], which reduces cholesterol concentrations in the plasma membrane [167]. Additionally, progesterone also increases the activity of 3-hydroxy-3-methylglutaryl (HMG-CoA) reductase, increasing the synthesis and membrane concentrations of the cholesterol precursors that are less active to support the high affinity oxytocin receptor [168]. The depletion of active membranous cholesterol forms leads to a low affinity mode of the oxytocin receptor,which may reduce its intracellular activity [169]. A decrease in the intracellular progesterone concentrations restores the cholesterol transport leading to an increase in active cholesterol concentration in the plasma membrane that supports the activity of high-affinity oxytocin receptor regaining its uterotonic effect [170].

Some of the activities of progesterone on myometrium may be mediated by its effects on the activity and the metabolism of cAMP [171] and inhibition of transmembrane Ca^{2+} entry [172]. Treatment of human myometrial smooth muscle cells with MPA resulted in a significant reduction in oxytocin-mediated increase in intracellular Ca^{2+} concentration [172].

3.3.5 The Interplay Between NFκB and Progesterone in Pregnancy Maintenance and in the Onset of Labor

Nuclear factor kappa B (NFκB) is a transcription factor family classically associated with inflammation. Data indicate that myometrial NFκB activity changes with labor and its activation is regulated in a spatio-temporal fashion. It has been proposed that NFκB is an upstream regulator of multiple labor-associated processes, including the formation of contraction-associated proteins, inflammatory mediators (e.g., cytokines), uterotonic phospholipid metabolites (e.g., PGs), and the induction of extracellular matrix remodeling [173, 174]. The stimuli and mechanisms responsible for NFκB activation in spontaneous labor have not yet been elucidated. Increasing local concentrations of surfactant protein A [175], accumulation of advance glycation end-products [176], amnion cells mechanical stretch [177], and the paracrine or autocrine pro-inflammatory effects of the corticotrophin-releasing hormone [178] have been proposed as potential candidates.

NFκB activation favors the myometrial expression of inhibitory isoforms of the PR. Evidence in support of this includes: 1) a spatial correlation is suggested by the enhanced expression of PR-B and PR-C along with NFκB activation during labor, and these changes are selective for the fundal human myometrium [36]; 2) a temporal correlation has been proposed due to the correlation of PR isoform expression and local NFκB activation in the pregnant mouse uterus and in human fundal myometrium [36, 175]; 3) intra-amniotic injection of surfactant protein A to pregnant mice, which promotes uterine NFκB activation and preterm labor, as well as a rapid increase in uterine levels of PR-B and PR-C [36, 175]; 4) intra-amniotic injection of NFκB inhibitor (SN50) caused a decrease in the uterine levels of PR-B and PR-C [36, 175]; and 5) *in vitro* models demonstrated that the activation of the NFκB pathway in response to IL-1β treatment is associated with an increased expression of all three PR isoforms (PR-A, PR-B and PR-C) in myometrial cells [36].

The PR-mediated activation of target genes that modulate uterine contractility is antagonized by NFκB. Kalkhoven et al. [179] reported the existence of a mutual trans-repression between the PR and the RelA(p65) subunit of NFκB in different cell lines. This repression was independent from the PR isoforms and the cell type. The authors suggested that the most likely mechanism involved is a direct interaction between the two proteins that would result in an inactive heterodimeric complex on the DNA, which prevents co-factors and members of the basal transcriptional machinery to initiate transcription [179]. Other possible explanations for the mutual repression of RelA(p65) and PR include binding of these transcription factors to their respective cognate DNA elements, or competition for the same co-activators or transcription intermediary factors (transcriptional interference or squelching) [179]. A similar mutual negative interaction between NFκB and PR activity was reported in human amnion cells [180]. Stimulation of these cells with IL-1β resulted in NFκB activation that was followed by repression of progesterone-dependent transcription, even in the presence of excess PR [180]. This may be the case during spontaneous labor in humans: indeed, the constitutive activity of NFκB reported in human amnion cells in the presence of labor may contribute to the loss of myometrial quiescence, both by repressing the PR activity and increasing the expression of COX-2. The authors proposed that the increase in NFκB activity, near to, or at the time of labor, may represent a "watershed point at which labor becomes inevitable" [180].

The anti-inflammatory activity of progesterone may contribute to the prolongation of pregnancy by direct or indirect attenuation of the NFκB-mediated inflammatory cascade. Several observations support this view: 1) over-expression of the PR in amnion cells was associated with significant repression of NFκB reporter expression [180]; 2) the IL-1β induced up-regulation of COX-2 mRNA in immortalized human fundal myometrial cells was suppressed by exogenous administration of progesterone and associated with a rapid induction of the NFκB transactivation inhibitor, kBα [181]; 3) progesterone down-regulates cytokine production by human leukemia cell lines, mediated, at least in part, by suppression of NFκB activity [182]; and 4) physiological concentrations of progesterone suppress both the spontaneous and the IL-1(α and β)-mediated production of IL-8 by the uterine cervical fibroblasts in pregnant rabbits [183].

In contrast, Vidaeff et al. [184] demonstrated that pre-treatment of HeLa cells with progesterone before exposure to IL-1β resulted in a significant decrease in NFκB protein subunit p65 in the cytoplasm. However, it did not reduce the amount of nuclear p65 or affect the nuclear translocation of p65. The authors suggested that any possible role played by progesterone in preterm labor prevention is not exerted through anti-inflammatory mechanisms of NFκB down-regulation [184].

3.3.6 Changes in Myometrial Progesterone Co-regulators During Pregnancy

The possibility that changes in the activity of co-regulators can contribute to the functional progesterone withdrawal is currently an object of investigation. Condon et al. [185] proposed that a decline in the levels of PR co-activators in the pregnant uterus at term may antagonize PR function and contribute to the initiation of labor. Analysis of the mRNA and protein expression of PR co-activators in the fundal myometrium of 12 women in labor and 12 women not in labor revealed that laboring myometrium was associated with a lower mRNA and protein expression of SRC-2, SRC-3 and CBP (the cAMP-response element-binding protein) than non-laboring myometrial samples, while SRC-1 expression was relatively unchanged before and after the onset of labor.

Term gestation was associated with a decrease in the levels of histone H3 acetylation in the human and mouse uterus. Treatment of pregnant mice with trichostatin, a histone deacetylase inhibitor, delayed the onset of labor by 24–48 h. Altogether, these results suggested that a reduced uterine expression of progesterone co-activators at term would lead to a reduction in histone acetylation resulting in an impaired PR responsiveness and a functional progesterone withdrawal [185].

In 2005, a novel progesterone co-repressor, polypyrimidine tract-binding protein-associated splicing factor (PSF), was identified in the rat myometrium [186]: its mRNA expression increased as term approached and was up-regulated prior the onset of labor. PSF interferes with PR binding to its DNA response element, and enhances PR degradation. Within the human myometrium, PSF expression was significantly up-regulated as pregnancy progressed, particularly within the upper uterine region, and levels remained elevated in labor. Co-immunoprecipitations and DNA-binding assays showed that PSF directly interacts with nuclear PR and glucocorticoid receptor and specific co-regulatory proteins within the human myometrium [187]. These findings are suggestive of a role for myometrial PSF as a nuclear co-regulator and a potential contributor to functional progesterone withdrawal [186–188].

3.3.7 Progesterone Receptor in Fetal Membranes, Decidua, Placenta

There is evidence that all three isoforms, PR-A, PR-B and PR-C, are expressed in decidua and fetal membranes, yet there is still controversy concerning the predominant isoform [189–193]. A Western Blot analysis conducted by Goldman et al. [189] revealed that the major isoform in the human decidua is PR-B, whereas in the human amnion it is PR-C. In contrast, in a review on this subject, Taylor et al. [192] reported that immunohistochemical, Western blotting and real time RT-PCR techniques provide evidence that the major PR isoform in human decidua is PR-A, whereas in human term fetal membranes and syncytiotrophoblast it is PR-C [192].

The quantitative and qualitative expression of the PR isoforms in the decidua and fetal membranes may be subject to significant changes during labor [189, 190]. Our group reported that in fetal membranes obtained from women in labor, there is a PR-A predominance and a higher PR-A/PR-B ratio than in women not in labor, in which PR-B is the predominant isoform [190]. Similarly, Mesiano et al. [152] reported that mRNA encoding for the PR-A and PR-A/PR-B expression ratio increased significantly in the human myometrium at term in association with labor. Although the role of progesterone receptors in the fetal membranes has not yet been elucidated, it has been proposed that a shift in progesterone isoform expression may be part of a "feto-maternal signaling pathway in the initiation of labor" [191].

3.3.8 The Effect of Progesterone on the Cervix During Pregnancy

Progesterone exerts biological effects in the uterine cervix [127–129, 194–201]; and a withdrawal (in rats, rabbits and sheep) or decline in progesterone action (guinea pigs and primates) [130] has been proposed as a key control mechanism for cervical ripening [128, 183, 201–207]. Evidence in support of this includes: 1) administration of anti-progestins to women in the mid-trimester and at term induces cervical ripening [205, 208–211] but not labor [128], that may not begin at all or may be delayed by days or weeks after cervical ripening has been accomplished; and 2) the administration of a progesterone-receptor antagonist to pregnant guinea pigs [203, 212], old-world monkeys [213] and *Tapaja Belangeri* induces cervical ripening [205]. Cervical responsiveness to anti-progestins increases with advancing gestational age, and the effect of anti-progestins in the cervix is not always accompanied by changes in myometrial activity [205]. Indeed, Stys et al. [214] demonstrated a dissociation between the effects of progesterone in the myometrium and those in the cervix. Contrary to the acute nature of uterine contractions, the process of cervical ripening is gradual in normal pregnancies, and may start weeks before labor and delivery [194, 199].

Cervical ripening is a multifactorial process affected by a myriad of factors [129, 215] and characterized by slow changes in the composition of the extra-cellular matrix. The precise mechanisms by which a blockade of progesterone action may induce cervical changes are poorly understood. Both *in vitro* and *in vivo* studies have supported the key role of progesterone in this process. A decline in progesterone action may induce cervical changes through pro-inflammatory mediators [200] including IL-8 [198], nitric oxide [204, 207], PGs [198] and matrix-degrading enzymes [197, 216, 217]. Cervical remodeling and ripening may be influenced by NFκB, which can oppose progesterone action [36, 134, 179–181, 184], providing a link between inflammation, a decline in progesterone action and cervical ripening. The effects of progesterone on cellular and extra-cellular components of the cervix are discussed below.

3.3.8.1 Collagen Remodeling

The mechanical properties of the cervix are largely determined by collagen and proteoglycans that comprise the extra-cellular matrix of the connective tissue. Shortly before labor, the cervix collagen fibers become less densely packed and the collagen concentration decreases [215, 218]. These changes are mediated, in part, by increased collagenase activity [219]. In pregnant women, there is a decrease of 30–50 % in the collagen concentration of the uterine cervix in comparison to the non-pregnant state [220]. Moreover, there is an association between high cervical collagen concentration and slow progress of labor [218].

Winn et al. [221] have investigated the individual and combined effects of relaxin, estrogen and progesterone on growth, softening and histological characteristics of the cervix of ovariectomized non-pregnant gilts. When administered alone, progesterone had no effect on cervical growth and only a modest effect on cervical softening; when progesterone was administrated with relaxin, there was an increased extensibility of the cervix in comparison to progesterone alone or to its combination with estrogen. In addition, the combination of progesterone and relaxin maximally decreased the collagen/amorphous ground substance [221]. The administration of mifepristone to pregnant rats at mid-gestation was associated with marked cervical changes, including decreased tensile strength, reduced collagen organization, and increased MMP-2 mRNA expression [222]. Additionally, the collagen fibrils in the cervix had a shorter mean length and smaller mean diameter after mifepristone treatment. Collectively, this evidence suggests that progesterone suppresses cervical collagenolysis, one of the major processes of cervical ripening before labor [222].

3.3.8.2 Changes in Glycosaminoglycans

Glycosaminoglycans (GAGs) are an important component of connective tissues and extracellular matrix. Softening of the cervix is associated with changes in GAG [194, 223], specifically, an increase in total GAGs, hyaluronic acid and water content, while sulphated GAGs decrease [194, 224–226]. Carbonne et al. [199] determined the effects of progesterone on PGE_2-induced changes in GAG synthesis in human cervical cells cultures. Progesterone did not prevent changes in GAG production (usually considered to reflect cervical ripening); moreover, high concentrations of this hormone even favored these changes. The authors hypothesized that this paradoxical finding may account for the early changes in the consistency of the cervix, and for the alteration in GAG content that can be observed as early as the first trimester [226, 227]. Another explanation for this counterintuitive finding is that progesterone has a different effect on the cervix and the body of the uterus. Increasing concentrations of progesterone during pregnancy may play a key role in the gradual ripening of the cervix, and promote the myometrial quiescence and down regulation of gap junctions in the uterus.

3.3.8.3 Suppression of Metalloproteinases

The degradation of collagen in the cervix is mediated primarily by MMPs, and their effect can be repressed by their endogenous inhibitors (TIMPs) [216, 228]. In the cervical fibroblast of rabbits, progesterone decreases the levels of proMMP-1, proMMP-3, and the steady-state levels of the respective mRNAs in the culture media, and increases the concentrations of TIMPs more effectively than that of estradiol-17 beta [197]. Similarly, Imada et al. [216] reported that physiological concentrations of progesterone suppressed IL-1 mediated production of proMMP-9 and its mRNA in a dose-dependent manner. The authors concluded that in the rabbit uterine cervix progesterone is a physiological suppressor of the proMMP-9 production at the transcriptional level [216]. However, the nature of the effect of progesterone may vary according to its concentrations: while high concentration of progesterone decreases production of collagenase in cervical fibroblasts of guinea pigs,a lower concentration of this hormone had the opposite effect [196].

The effect of anti-progestins on cervical MMP expression has also been studied. Onapristone augmented the expression of MMP-3 mRNA in rabbits [217]. In contrast, mifepristone increases the expression of MMP-2 mRNA, but not of MMP-9 or

MMP-3 [222], suggesting that the anti-progestins differ not only in their specificity to progesterone but they may also differ in the mechanism by which their effect is exerted.

3.3.8.4 Modulation of the Inflammatory Response in the Uterine Cervix

Macrophages, neutrophils and eosinophils are thought to play a central role in the remodeling of the cervical connective tissue by production of cytokines and proteolytic enzymes in response to inflammatory stimuli, and have a regulatory role in cervical ripening [229, 230]. Ramos et al. [231] investigated the mechanism through which eosinophilic invasion is modulated during the second half of pregnancy in rats. Exposure to 17β-estradiol together with progesterone resulted in very poor eosinophilic infiltration, but the progesterone inhibition of eosinophilic infiltration was reversed by co-administration of mifepristone. The authors suggested that the progesterone effect is mediated through the progestin receptor [231].

Human cervical cells release IL-8 [232], a neutrophilic chemotactic and activating agent [233], which is thought to initiate cervical ripening by promoting neutrophil chemotaxis to the cervix and their activation within the cervical stroma [232, 233]. Denison et al. [198] demonstrated that the release of IL-8 by cervical explants was significantly stimulated by PGE_2 and inhibited by progesterone. The release of the secretory leukocyte protease inhibitor (an inhibitor of neutrophil function) by cervical explants was significantly stimulated by progesterone and inhibited by PGE_2 [198]. Moreover, *in vivo* study in mice found that the administration of MPA before intrauterine lipopolysaccharide significantly increases the expression of the anti-inflammatory cytokine IL-10, and can inhibit inflammation-induced preterm labor [200].

4 Conclusions

Progesterone is the key hormone in pregnancy maintenance and is involved in all processes from the preparation of the uterine decidua, myometrium and cervix during the menstrual cycle through blastocyst implantation, sustaining of myometrial quiescence, cervical competence and modulation of the maternal immune system during pregnancy. There is accumulating evidence that progesterone withdrawal during parturition in humans is probably functional and involves a shift in the balance between progesterone and cortisol, as well as changes in the genomic and non-genomic effects of progesterone at the cellular level.

Acknowledgment This chapter was supported by the Intramural Research Program of the National Institute of Child Health and Human Development, NIH, DHHS.

References

1. Asdell SA. The growth and function of the corpus luteum. Physiol Rev. 1928;8:313–45.
2. Malpighi M. Opera omnia, seu thesaurus locupletissimus botanico-medico-anatomicus. Leiden: P. van der Aa; 1687
3. Fraenkel L. Die funktion des corpus luteum. Definition der theorie, ihre entstehung. Archiv Gynaekol. 1903;68:438–43.
4. Corner GW, Allen WM. Physiology of the corpus luteum, II: production of a special uterine reaction (progestational proliferation) by extracts of the corpus luteum. Am J Physiol. 1929;88:326–99.
5. Allen WM, Corner GW. Physiology of the corpus luteum, III: normal growth and implantation of embryos after very early ablation of the ovaries, and under the influence of extracts of the corpus luteum. Am J Physiol. 1929;88:340–6.
6. Allen WM. Physiology of the corpus luteum, V: the preparation and some chemical properties of progestin, a hormone of the corpus luteum which produces progestational proliferation. Am J Physiol. 1930;92:174–88.
7. Wintersteiner O, Allen WM. Crystalline progestin. J Biol Chem. 1934;107:321–36.
8. Slotta K, Ruschig H, Fels E. Reindarstellung der Hormone aus dem Corpus Luteum. Ber Dtsch Chem Ges. 1934;67:1270–3.
9. Hartmann M, Wettstein A. Ein krystallisiertes Hormon aus Corpus luteum. Helv Chim Acta. 1934;17:878–82.
10. Butenandt A, Westphal U. Zur Isolierung und Characterisierung des Corpus-luteum-Hormons. Ber Dtsch Chem Ges. 1934;67:1440–2.
11. Allen WM. Nomenclature of the corpus luteum hormone. Science. 1935;82:153.
12. Graham JD, Clarke CL. Physiological action of progesterone in target tissues. Endocr Rev. 1997;18(4):502–19.
13. Petersen SL, et al. Novel progesterone receptors: neural localization and possible functions. Front Neurosci. 2013;7:164.
14. Singh M, Su C, Ng S. Non-genomic mechanisms of progesterone action in the brain. Front Neurosci. 2013;7:159.
15. Nath A, Sitruk-Ware R. Different cardiovascular effects of progestins according to structure and activity. Climacteric. 2009;12 Suppl 1:96–101.
16. Kaore SN, et al. Novel actions of progesterone: what we know today and what will be the scenario in the future? J Pharm Pharmacol. 2012;64(8):1040–62.
17. DeMayo FJ, et al. Mechanisms of action of estrogen and progesterone. Ann N Y Acad Sci. 2002;955:48–59.
18. Catt KJ. IV. Reproductive endocrinology. Lancet. 1970;1(7656):1097–104.
19. An BS, et al. Differential role of progesterone receptor isoforms in the transcriptional regulation of human gonadotropin-releasing hormone I (GnRH I) receptor, GnRH I, and GnRH II. J Clin Endocrinol Metab. 2005;90(2):1106–13.
20. Williams SP, Sigler PB. Atomic structure of progesterone complexed with its receptor. Nature. 1998;393(6683):392–6.
21. Losel R, Wehling M. Nongenomic actions of steroid hormones. Nat Rev Mol Cell Biol. 2003;4(1):46–56.
22. Losel RM, et al. Nongenomic steroid action: controversies, questions, and answers. Physiol Rev. 2003;83(3):965–1016.
23. Tsai MJ, O'Malley BW. Molecular mechanisms of action of steroid/thyroid receptor superfamily members. Annu Rev Biochem. 1994;63:451–86.
24. Power RF, Conneely OM, O'Malley BW. New insights into activation of the steroid hormone receptor superfamily. Trends Pharmacol Sci. 1992;13(8):318–23.

25. DeMarzo AM, et al. Dimerization of mammalian progesterone receptors occurs in the absence of DNA and is related to the release of the 90-kDa heat shock protein. Proc Natl Acad Sci U S A. 1991;88(1):72–6.
26. Brosens JJ, et al. Steroid receptor action. Best Pract Res Clin Obstet Gynaecol. 2004;18(2): 265–83.
27. Kastner P, et al. Two distinct estrogen-regulated promoters generate transcripts encoding the two functionally different human progesterone receptor forms A and B. EMBO J. 1990;9(5):1603–14.
28. Meyer ME, et al. A limiting factor mediates the differential activation of promoters by the human progesterone receptor isoforms. J Biol Chem. 1992;267(15):10882–7.
29. Vegeto E, et al. Human progesterone receptor A form is a cell- and promoter-specific repressor of human progesterone receptor B function. Mol Endocrinol. 1993;7(10):1244–55.
30. Hirata S, et al. Isoform/variant mRNAs for sex steroid hormone receptors in humans. Trends Endocrinol Metab. 2003;14(3):124–9.
31. Sartorius CA, et al. A third transactivation function (AF3) of human progesterone receptors located in the unique N-terminal segment of the B-isoform. Mol Endocrinol. 1994;8(10): 1347–60.
32. Huse B, et al. Definition of a negative modulation domain in the human progesterone receptor. Mol Endocrinol. 1998;12(9):1334–42.
33. Wildman DE, et al. Evolutionary history of the progesterone receptor in primates. J Soc Gynecol Investig. 2006;13(2):238A.
34. Wei LL, et al. 5′-Heterogeneity in human progesterone receptor transcripts predicts a new amino-terminal truncated "C"-receptor and unique A-receptor messages. Mol Endocrinol. 1990;4(12):1833–40.
35. Wei LL, et al. An amino-terminal truncated progesterone receptor isoform, PRc, enhances progestin-induced transcriptional activity. Mol Endocrinol. 1996;10(11):1379–87.
36. Condon JC, et al. Up-regulation of the progesterone receptor (PR)-C isoform in laboring myometrium by activation of nuclear factor-kappaB may contribute to the onset of labor through inhibition of PR function. Mol Endocrinol. 2006;20(4):764–75.
37. Wei LL, Norris BM, Baker CJ. An N-terminally truncated third progesterone receptor protein, PR(C), forms heterodimers with PR(B) but interferes in PR(B)-DNA binding. J Steroid Biochem Mol Biol. 1997;62(4):287–97.
38. Hirata S, et al. The novel isoform of the estrogen receptor-alpha cDNA (ERalpha isoform S cDNA) in the human testis. J Steroid Biochem Mol Biol. 2002;80(3):299–305.
39. Saner KJ, et al. Cloning and expression of a novel, truncated, progesterone receptor. Mol Cell Endocrinol. 2003;200(1–2):155–63.
40. Samalecos A, Gellersen B. Systematic expression analysis and antibody screening do not support the existence of naturally occurring progesterone receptor (PR)-C, PR-M, or other truncated PR isoforms. Endocrinology. 2008;149(11):5872–87.
41. Madsen G, et al. Progesterone receptor or cytoskeletal protein? Reprod Sci. 2007;14(3): 217–22.
42. Kumar R, et al. The clinical relevance of steroid hormone receptor corepressors. Clin Cancer Res. 2005;11(8):2822–31.
43. Lee K, et al. Molecular mechanisms involved in progesterone receptor regulation of uterine function. J Steroid Biochem Mol Biol. 2006;102(1–5):41–50.
44. Spelsberg TC, Steggles AW, O'Malley BW. Progesterone-binding components of chick oviduct. 3. Chromatin acceptor sites. J Biol Chem. 1971;246(13):4188–97.
45. Gao X, Loggie BW, Nawaz Z. The roles of sex steroid receptor coregulators in cancer. Mol Cancer. 2002;1:7.
46. Mukherjee A, et al. Steroid receptor coactivator 2 is essential for progesterone-dependent uterine function and mammary morphogenesis: insights from the mouse–implications for the human. J Steroid Biochem Mol Biol. 2006;102(1–5):22–31.
47. Fernandez-Valdivia R, et al. Progesterone-action in the murine uterus and mammary gland requires steroid receptor coactivator 2: relevance to the human. Front Biosci. 2007;12:3640–7.

48. McKenna NJ, O'Malley BW. Combinatorial control of gene expression by nuclear receptors and coregulators. Cell. 2002;108(4):465–74.
49. Smith CL, O'Malley BW. Coregulator function: a key to understanding tissue specificity of selective receptor modulators. Endocr Rev. 2004;25(1):45–71.
50. Xu J, et al. Partial hormone resistance in mice with disruption of the steroid receptor coactivator-1 (SRC-1) gene. Science. 1998;279(5358):1922–5.
51. Han SJ, et al. Steroid receptor coactivator (SRC)-1 and SRC-3 differentially modulate tissue-specific activation functions of the progesterone receptor. Mol Endocrinol. 2006;20(1):45–55.
52. Xu J, et al. The steroid receptor coactivator SRC-3 (p/CIP/RAC3/AIB1/ACTR/TRAM-1) is required for normal growth, puberty, female reproductive function, and mammary gland development. Proc Natl Acad Sci U S A. 2000;97(12):6379–84.
53. Heery DM, et al. A signature motif in transcriptional co-activators mediates binding to nuclear receptors. Nature. 1997;387(6634):733–6.
54. Aoyagi S, Archer TK. Dynamic histone acetylation/deacetylation with progesterone receptor-mediated transcription. Mol Endocrinol. 2007;21(4):843–56.
55. Blackmore PF, Lattanzio FA. Cell surface localization of a novel non-genomic progesterone receptor on the head of human sperm. Biochem Biophys Res Commun. 1991;181(1):331–6.
56. Baldi E, et al. Intracellular calcium accumulation and responsiveness to progesterone in capacitating human spermatozoa. J Androl. 1991;12(5):323–30.
57. Luconi M, et al. Identification and characterization of functional nongenomic progesterone receptors on human sperm membrane. J Clin Endocrinol Metab. 1998;83(3):877–85.
58. Falkenstein E, et al. Specific progesterone binding to a membrane protein and related nongenomic effects on Ca2+-fluxes in sperm. Endocrinology. 1999;140(12):5999–6002.
59. Patrat C, Serres C, Jouannet P. Induction of a sodium ion influx by progesterone in human spermatozoa. Biol Reprod. 2000;62(5):1380–6.
60. Turner KO, Meizel S. Progesterone-mediated efflux of cytosolic chloride during the human sperm acrosome reaction. Biochem Biophys Res Commun. 1995;213(3):774–80.
61. Finidori-Lepicard J, et al. Progesterone inhibits membrane-bound adenylate cyclase in Xenopus laevis oocytes. Nature. 1981;292(5820):255–7.
62. Grosse B, et al. Membrane signalling and progesterone in female and male osteoblasts. I. Involvement of intracellular Ca(2+), inositol trisphosphate, and diacylglycerol, but not cAMP. J Cell Biochem. 2000;79(2):334–45.
63. Le Mellay V, Lieberherr M. Membrane signaling and progesterone in female and male osteoblasts. II. Direct involvement of G alpha q/11 coupled to PLC-beta 1 and PLC-beta 3. J Cell Biochem. 2000;79(2):173–81.
64. Maller JL, Krebs EG. Progesterone-stimulated meiotic cell division in Xenopus oocytes. Induction by regulatory subunit and inhibition by catalytic subunit of adenosine 3′:5′-monophosphate-dependent protein kinase. J Biol Chem. 1977;252(5):1712–8.
65. Ishikawa K, et al. Primary action of steroid hormone at the surface of amphibian oocyte in the induction of germinal vesicle breakdown. Mol Cell Endocrinol. 1977;9(1):91–100.
66. Baulieu EE, et al. Steroid-induced meiotic division in Xenopus laevis oocytes: surface and calcium. Nature. 1978;275(5681):593–8.
67. Meizel S, Turner KO. Progesterone acts at the plasma membrane of human sperm. Mol Cell Endocrinol. 1991;77(1–3):R1–5.
68. Meyer C, et al. Purification and partial sequencing of high-affinity progesterone-binding site(s) from porcine liver membranes. Eur J Biochem. 1996;239(3):726–31.
69. Falkenstein E, et al. Full-length cDNA sequence of a progesterone membrane-binding protein from porcine vascular smooth muscle cells. Biochem Biophys Res Commun. 1996;229(1):86–9.
70. Krebs CJ, et al. A membrane-associated progesterone-binding protein, 25-Dx, is regulated by progesterone in brain regions involved in female reproductive behaviors. Proc Natl Acad Sci U S A. 2000;97(23):12816–21.

71. Zhu Y, Bond J, Thomas P. Identification, classification, and partial characterization of genes in humans and other vertebrates homologous to a fish membrane progestin receptor. Proc Natl Acad Sci U S A. 2003;100(5):2237–42.
72. White HD, et al. Mucosal immunity in the human female reproductive tract: cytotoxic T lymphocyte function in the cervix and vagina of premenopausal and postmenopausal women. Am J Reprod Immunol. 1997;37(1):30–8.
73. Wira C, et al. Effect of the menstrual cycle on immunological parameters in the human female reproductive tract. J Acquir Immune Defic Syndr. 2005;38 Suppl 1:S34–6.
74. Wira CR, Rossoll RM. Antigen-presenting cells in the female reproductive tract: influence of sex hormones on antigen presentation in the vagina. Immunology. 1995;84(4):505–8.
75. Walch KT, Huber JC. Progesterone for recurrent miscarriage: truth and deceptions. Best Pract Res Clin Obstet Gynaecol. 2008;22(2):375–89.
76. De M, Wood GW. Influence of oestrogen and progesterone on macrophage distribution in the mouse uterus. J Endocrinol. 1990;126(3):417–24.
77. Hanna J, et al. Decidual NK cells regulate key developmental processes at the human fetal-maternal interface. Nat Med. 2006;12(9):1065–74.
78. Croy BA, et al. Decidual natural killer cells: key regulators of placental development (a review). J Reprod Immunol. 2002;57(1–2):151–68.
79. Beagley KW, Gockel CM. Regulation of innate and adaptive immunity by the female sex hormones oestradiol and progesterone. FEMS Immunol Med Microbiol. 2003;38(1):13–22.
80. Henderson TA, et al. Steroid receptor expression in uterine natural killer cells. J Clin Endocrinol Metab. 2003;88(1):440–9.
81. Verma S, et al. Human decidual natural killer cells express the receptor for and respond to the cytokine interleukin 15. Biol Reprod. 2000;62(4):959–68.
82. Roussev RG, Higgins NG, McIntyre JA. Phenotypic characterization of normal human placental mononuclear cells. J Reprod Immunol. 1993;25(1):15–29.
83. Chao KH, et al. Decidual natural killer cytotoxicity decreased in normal pregnancy but not in anembryonic pregnancy and recurrent spontaneous abortion. Am J Reprod Immunol. 1995;34(5):274–80.
84. Piccinni MP, Maggi E, Romagnani S. Role of hormone-controlled T-cell cytokines in the maintenance of pregnancy. Biochem Soc Trans. 2000;28(2):212–5.
85. Szekeres-Bartho J, Wegmann TG. A progesterone-dependent immunomodulatory protein alters the Th1/Th2 balance. J Reprod Immunol. 1996;31(1–2):81–95.
86. Saito S. Cytokine network at the feto-maternal interface. J Reprod Immunol. 2000;47(2):87–103.
87. Eblen AC, et al. Alterations in humoral immune responses associated with recurrent pregnancy loss. Fertil Steril. 2000;73(2):305–13.
88. Druckmann R, Druckmann MA. Progesterone and the immunology of pregnancy. J Steroid Biochem Mol Biol. 2005;97(5):389–96.
89. Szekeres-Bartho J, et al. The mechanism of the inhibitory effect of progesterone on lymphocyte cytotoxicity: I. Progesterone-treated lymphocytes release a substance inhibiting cytotoxicity and prostaglandin synthesis. Am J Reprod Immunol Microbiol. 1985;9(1):15–8.
90. Kelemen K, et al. A progesterone-induced protein increases the synthesis of asymmetric antibodies. Cell Immunol. 1996;167(1):129–34.
91. Faust Z, et al. Progesterone-induced blocking factor inhibits degranulation of natural killer cells. Am J Reprod Immunol. 1999;42(2):71–5.
92. Laskarin G, et al. Progesterone induced blocking factor (PIBF) mediates progesterone induced suppression of decidual lymphocyte cytotoxicity. Am J Reprod Immunol. 2002;48(4):201–9.
93. Jabbour HN, et al. Endocrine regulation of menstruation. Endocr Rev. 2006;27(1):17–46.
94. Gambino LS, et al. Angiogenesis occurs by vessel elongation in proliferative phase human endometrium. Hum Reprod. 2002;17(5):1199–206.
95. Noyes RW, Hertig AT, Rock J. Dating the endometrial biopsy. Am J Obstet Gynecol. 1975;122(2):262–3.

96. Lerner LJ. Hormone antagonists: inhibitors of specific activities of estrogen and androgen. Recent Prog Horm Res. 1964;20:435–90.
97. Hsueh AJ, Peck Jr EJ, Clark JH. Progesterone antagonism of the oestrogen receptor and oestrogen-induced uterine growth. Nature. 1975;254(5498):337–9.
98. Tseng L, Gurpide E. Induction of human endometrial estradiol dehydrogenase by progestins. Endocrinology. 1975;97(4):825–33.
99. Kirkland JL, Murthy L, Stancel GM. Progesterone inhibits the estrogen-induced expression of c-fos messenger ribonucleic acid in the uterus. Endocrinology. 1992;130(6):3223–30.
100. Lockwood CJ, et al. The role of progestationally regulated stromal cell tissue factor and type-1 plasminogen activator inhibitor (PAI-1) in endometrial hemostasis and menstruation. Ann N Y Acad Sci. 1994;734:57–79.
101. Lockwood CJ, et al. Biological mechanisms underlying RU 486 clinical effects: inhibition of endometrial stromal cell tissue factor content. J Clin Endocrinol Metab. 1994;79(3):786–90.
102. Cibils LA. Contractility of the nonpregnant human uterus. Obstet Gynecol. 1967;30(3): 441–61.
103. Bulletti C, et al. Uterine contractility during the menstrual cycle. Hum Reprod. 2000;15 Suppl 1:81–9.
104. de Ziegler D, et al. Contractility of the nonpregnant uterus: the follicular phase. Ann N Y Acad Sci. 2001;943:172–84.
105. Noe M, et al. The cyclic pattern of the immunocytochemical expression of oestrogen and progesterone receptors in human myometrial and endometrial layers: characterization of the endometrial-subendometrial unit. Hum Reprod. 1999;14(1):190–7.
106. Akerlund M, Batra S, Helm G. Comparison of plasma and myometrial tissue concentrations of estradiol-17 beta and progesterone in nonpregnant women. Contraception. 1981;23(4): 447–55.
107. Batra S, Sjoberg NO, Thorbert G. Sex steroids in plasma and reproductive tissues of the female guinea pig. Biol Reprod. 1980;22(3):430–7.
108. Jordan J, Albert S. The Cervix. 2nd ed. Oxford, United Kingdom: Blackwell Publising; 2007.
109. Cano A, et al. Expression of estrogen receptors, progesterone receptors, and an estrogen receptor-associated protein in the human cervix during the menstrual cycle and menopause. Fertil Steril. 1990;54(6):1058–64.
110. Gorodeski GI. Effects of menopause and estrogen on cervical epithelial permeability. J Clin Endocrinol Metab. 2000;85(7):2584–95.
111. Odeblad E. Physical properties of cervical mucus. Adv Exp Med Biol. 1977;89:217–25.
112. Snijders MP, et al. Immunocytochemical analysis of oestrogen receptors and progesterone receptors in the human uterus throughout the menstrual cycle and after the menopause. J Reprod Fertil. 1992;94(2):363–71.
113. Odeblad E. The physics of the cervical mucus. Acta Obstet Gynecol Scand Suppl. 1959;38(Supp 1):44–58.
114. Odeblad E. Undulations of macromolecules in cervical mucus. Int J Fertil. 1962;7:313–9.
115. Odeblad E, WESTIN B. Studies on the penetration of radioactive ions through human cervical mucus. Acta Radiol. 1958;49(5):382–8.
116. Croxatto HB. Mechanisms that explain the contraceptive action of progestin implants for women. Contraception. 2002;65(1):21–7.
117. Erkkola R, Landgren BM. Role of progestins in contraception. Acta Obstet Gynecol Scand. 2005;84(3):207–16.
118. Mesiano S. Roles of estrogen and progesterone in human parturition. Front Horm Res. 2001;27:86–104.
119. Tulchinsky D, Hobel CJ. Plasma human chorionic gonadotropin, estrone, estradiol, estriol, progesterone, and 17 alpha-hydroxyprogesterone in human pregnancy. 3. Early normal pregnancy. Am J Obstet Gynecol. 1973;117(7):884–93.
120. Johansson ED. Plasma levels of progesterone in pregnancy measured by a rapid competitive protein binding technique. Acta Endocrinol (Copenh). 1969;61(4):607–17.

121. Csapo AI, Pulkkinen MO, Wiest WG. Effects of luteectomy and progesterone replacement therapy in early pregnant patients. Am J Obstet Gynecol. 1973;115(6):759–65.
122. Tulchinsky D, Okada D. Hormones in human pregnancy. IV Plasma progesterone. Am J Obstet Gynecol. 1975;121(3):293–9.
123. Sippell WG, et al. Concentrations of aldosterone, corticosterone, 11-deoxycorticosterone, progesterone, 17-hydroxyprogesterone, 11-deoxycortisol, cortisol, and cortisone determined simultaneously in human amniotic fluid throughout gestation. J Clin Endocrinol Metab. 1981;52(3):385–92.
124. Nagamani M, et al. Maternal and amniotic fluid steroids throughout human pregnancy. Am J Obstet Gynecol. 1979;134(6):674–80.
125. Ohana E, et al. Maternal plasma and amniotic fluid cortisol and progesterone concentrations between women with and without term labor. A comparison. J Reprod Med. 1996;41(2):80–6.
126. Mazor M, et al. Maternal plasma and amniotic fluid 17 beta-estradiol, progesterone and cortisol concentrations in women with successfully and unsuccessfully treated preterm labor. Arch Gynecol Obstet. 1996;258(2):89–96.
127. Gorodeski IG, et al. Progesterone (P) receptor dynamics in estrogen primed normal human cervix following P injection. Fertil Steril. 1987;47(1):108–13.
128. Chwalisz K. The use of progesterone antagonists for cervical ripening and as an adjunct to labour and delivery. Hum Reprod. 1994;9 Suppl 1:131–61.
129. Stjernholm Y, et al. Cervical ripening in humans: potential roles of estrogen, progesterone, and insulin-like growth factor-I. Am J Obstet Gynecol. 1996;174(3):1065–71.
130. Bernal AL. Overview of current research in parturition. Exp Physiol. 2001;86(2):213–22.
131. Karim SM, Hillier K. Prostaglandins in the control of animal and human reproduction. Br Med Bull. 1979;35(2):173–80.
132. Zakar T, Hertelendy F. Progesterone withdrawal: key to parturition. Am J Obstet Gynecol. 2007;196(4):289–96.
133. Thorburn GD, Challis J, Currie WB. Control of parturition in domestic animals. Biol Reprod. 1977;16(1):18–27.
134. Mendelson CR, Condon JC. New insights into the molecular endocrinology of parturition. J Steroid Biochem Mol Biol. 2005;93(2–5):113–9.
135. Mahendroo MS, et al. The parturition defect in steroid 5alpha-reductase type 1 knockout mice is due to impaired cervical ripening. Mol Endocrinol. 1999;13(6):981–92.
136. Bygdeman M, et al. The use of progesterone antagonists in combination with prostaglandin for termination of pregnancy. Hum Reprod. 1994;9 Suppl 1:121–5.
137. Puri CP, et al. Effects of progesterone antagonist ZK 98.299 on early pregnancy and foetal outcome in bonnet monkeys. Contraception. 1990;41(2):197–205.
138. Westphal U, Stroupe SD, Cheng SL. Progesterone binding to serum proteins. Ann N Y Acad Sci. 1977;286:10–28.
139. Karalis K, Goodwin G, Majzoub JA. Cortisol blockade of progesterone: a possible molecular mechanism involved in the initiation of human labor. Nat Med. 1996;2(5):556–60.
140. Milewich L, et al. Initiation of human parturition. VIII. Metabolism of progesterone by fetal membranes of early and late human gestation. Obstet Gynecol. 1977;50(1):45–8.
141. Mitchell BF, Wong S. Changes in 17 beta,20 alpha-hydroxysteroid dehydrogenase activity supporting an increase in the estrogen/progesterone ratio of human fetal membranes at parturition. Am J Obstet Gynecol. 1993;168(5):1377–85.
142. Pieber D, et al. Interactions between progesterone receptor isoforms in myometrial cells in human labour. Mol Hum Reprod. 2001;7(9):875–9.
143. Rezapour M, et al. Sex steroid receptors and human parturition. Obstet Gynecol. 1997;89(6):918–24.
144. How H, et al. Myometrial estradiol and progesterone receptor changes in preterm and term pregnancies. Obstet Gynecol. 1995;86(6):936–40.
145. Tan H, et al. Progesterone receptor-A and -B have opposite effects on proinflammatory gene expression in human myometrial cells: implications for progesterone actions in human pregnancy and parturition. J Clin Endocrinol Metab. 2012;97(5):E719–30.

146. Challis JRG, et al. Endocrine and paracrine regulation of birth at term and preterm. Endocr Rev. 2000;21(5):514–50.
147. Fu X, et al. Unexpected stimulatory effect of progesterone on human myometrial contractile activity in vitro. Obstet Gynecol. 1993;82(1):23–8.
148. Fu X, et al. Antitachyphylactic effects of progesterone and oxytocin on term human myometrial contractile activity in vitro. Obstet Gynecol. 1993;82(4 Pt 1):532–8.
149. Mesiano S. Myometrial progesterone responsiveness. Semin Reprod Med. 2007;25(1): 5–13.
150. Merlino AA, et al. Nuclear progesterone receptors in the human pregnancy myometrium: evidence that parturition involves functional progesterone withdrawal mediated by increased expression of progesterone receptor-A. J Clin Endocrinol Metab. 2007;92(5):1927–33.
151. Pieber D, Allport VC, Bennett PR. Progesterone receptor isoform A inhibits isoform B-mediated transactivation in human amnion. Eur J Pharmacol. 2001;427(1):7–11.
152. Mesiano S, et al. Progesterone withdrawal and estrogen activation in human parturition are coordinated by progesterone receptor A expression in the myometrium. J Clin Endocrinol Metab. 2002;87(6):2924–30.
153. Haluska GJ, et al. Progesterone receptor localization and isoforms in myometrium, decidua, and fetal membranes from rhesus macaques: evidence for functional progesterone withdrawal at parturition. J Soc Gynecol Investig. 2002;9(3):125–36.
154. Madsen G, et al. Prostaglandins differentially modulate progesterone receptor-A and -B expression in human myometrial cells: evidence for prostaglandin-induced functional progesterone withdrawal. J Clin Endocrinol Metab. 2004;89(2):1010–3.
155. Karteris E, et al. Progesterone signaling in human myometrium through two novel membrane G protein-coupled receptors: potential role in functional progesterone withdrawal at term. Mol Endocrinol. 2006;20(7):1519–34.
156. Haukkamaa M. High affinity progesterone binding sites of human uterine microsomal membranes. J Steroid Biochem. 1984;20(2):569–73.
157. Fernandes MS, et al. Regulated expression of putative membrane progestin receptor homologues in human endometrium and gestational tissues. J Endocrinol. 2005;187(1):89–101.
158. Nissenson R, Fluoret G, Hechter O. Opposing effects of estradiol and progesterone on oxytocin receptors in rabbit uterus. Proc Natl Acad Sci U S A. 1978;75(4):2044–8.
159. Soloff MS, et al. Regulation of oxytocin receptor concentration in rat uterine explants by estrogen and progesterone. Can J Biochem Cell Biol. 1983;61(7):625–30.
160. Larcher A, et al. Oxytocin receptor gene expression in the rat uterus during pregnancy and the estrous cycle and in response to gonadal steroid treatment. Endocrinology. 1995;136(12): 5350–6.
161. KNAUS H. Zur Physiologie des Corpus luteum. Arch Gynak. 1929;138(1):201–16.
162. Grazzini E, et al. Inhibition of oxytocin receptor function by direct binding of progesterone. Nature. 1998;392(6675):509–12.
163. Astle S, Khan RN, Thornton S. The effects of a progesterone metabolite, 5 beta-dihydroprogesterone, on oxytocin receptor binding in human myometrial membranes. BJOG. 2003;110(6):589–92.
164. Gimpl G, Fahrenholz F. The oxytocin receptor system: structure, function, and regulation. Physiol Rev. 2001;81(2):629–83.
165. Gimpl G, et al. Oxytocin receptors and cholesterol: interaction and regulation. Exp Physiol 2000;85 Spec No:41S–49S
166. Debry P, et al. Role of multidrug resistance P-glycoproteins in cholesterol esterification. J Biol Chem. 1997;272(2):1026–31.
167. Smart EJ, et al. A role for caveolin in transport of cholesterol from endoplasmic reticulum to plasma membrane. J Biol Chem. 1996;271(46):29427–35.
168. Metherall JE, Waugh K, Li H. Progesterone inhibits cholesterol biosynthesis in cultured cells. Accumulation of cholesterol precursors. J Biol Chem. 1996;271(5):2627–33.
169. Gimpl G, Fahrenholz F. Human oxytocin receptors in cholesterol-rich vs. cholesterol-poor microdomains of the plasma membrane. Eur J Biochem. 2000;267(9):2483–97.

170. Klein U, Gimpl G, Fahrenholz F. Alteration of the myometrial plasma membrane cholesterol content with beta-cyclodextrin modulates the binding affinity of the oxytocin receptor. Biochemistry. 1995;34(42):13784–93.
171. Kofinas AD, et al. Progesterone and estradiol concentrations in nonpregnant and pregnant human myometrium. Effect of progesterone and estradiol on cyclic adenosine monophosphate-phosphodiesterase activity. J Reprod Med. 1990;35(11):1045–50.
172. Fomin VP, Cox BE, Word RA. Effect of progesterone on intracellular Ca2+ homeostasis in human myometrial smooth muscle cells. Am J Physiol. 1999;276(2 Pt 1):C379–85.
173. Lindstrom TM, Bennett PR. The role of nuclear factor kappa B in human labour. Reproduction. 2005;130(5):569–81.
174. Lappas M, Rice GE. The role and regulation of the nuclear factor kappa B signalling pathway in human labour. Placenta. 2007;28(5–6):543–56.
175. Condon JC, et al. Surfactant protein secreted by the maturing mouse fetal lung acts as a hormone that signals the initiation of parturition. Proc Natl Acad Sci U S A. 2004;101(14):4978–83.
176. Lappas M, Permezel M, Rice GE. Advanced glycation end products mediate pro-inflammatory actions in human gestational tissues via nuclear factor-kappaB and extracellular signal-regulated kinase 1/2. J Endocrinol. 2007;193(2):269–77.
177. Mohan AR, et al. The effect of mechanical stretch on cyclooxygenase type 2 expression and activator protein-1 and nuclear factor-kappaB activity in human amnion cells. Endocrinology. 2007;148(4):1850–7.
178. Karalis K, et al. Autocrine or paracrine inflammatory actions of corticotropin-releasing hormone in vivo. Science. 1991;254(5030):421–3.
179. Kalkhoven E, et al. Negative interaction between the RelA(p65) subunit of NF-kappaB and the progesterone receptor. J Biol Chem. 1996;271(11):6217–24.
180. Allport VC, et al. Human labour is associated with nuclear factor-kappaB activity which mediates cyclo-oxygenase-2 expression and is involved with the 'functional progesterone withdrawal'. Mol Hum Reprod. 2001;7(6):581–6.
181. Hardy DB, et al. Progesterone receptor plays a major antiinflammatory role in human myometrial cells by antagonism of nuclear factor-kappaB activation of cyclooxygenase 2 expression. Mol Endocrinol. 2006;20(11):2724–33.
182. Srivastava MD, Anderson DJ. Progesterone receptor expression by human leukocyte cell lines: molecular mechanisms of cytokine suppression. Clin Exp Obstet Gynecol. 2007;34(1):14–24.
183. Ito A, et al. Suppression of interleukin 8 production by progesterone in rabbit uterine cervix. Biochem J. 1994;301(Pt 1):183–6.
184. Vidaeff AC, et al. Impact of progesterone on cytokine-stimulated nuclear factor-kappaB signaling in HeLa cells. J Matern Fetal Neonatal Med. 2007;20(1):23–8.
185. Condon JC, et al. A decline in the levels of progesterone receptor coactivators in the pregnant uterus at term may antagonize progesterone receptor function and contribute to the initiation of parturition. Proc Natl Acad Sci U S A. 2003;100(16):9518–23.
186. Dong X, et al. Identification and characterization of the protein-associated splicing factor as a negative co-regulator of the progesterone receptor. J Biol Chem. 2005;280(14):13329–40.
187. Tyson-Capper AJ, Shiells EA, Robson SC. Interplay between polypyrimidine tract binding protein-associated splicing factor and human myometrial progesterone receptors. J Mol Endocrinol. 2009;43(1):29–41.
188. Xie N, et al. Expression and function of myometrial PSF suggest a role in progesterone withdrawal and the initiation of labor. Mol Endocrinol. 2012;26(8):1370–9.
189. Goldman S, et al. Progesterone receptor expression in human decidua and fetal membranes before and after contractions: possible mechanism for functional progesterone withdrawal. Mol Hum Reprod. 2005;11(4):269–77.
190. Oh SY, et al. Progesterone receptor isoform (A/B) ratio of human fetal membranes increases during term parturition. Am J Obstet Gynecol. 2005;193(3 Pt 2):1156–60.

191. Mills AA, et al. Characterization of progesterone receptor isoform expression in fetal membranes. Am J Obstet Gynecol. 2006;195(4):998–1003.
192. Taylor AH, et al. The progesterone receptor in human term amniochorion and placenta is isoform C. Endocrinology. 2006;147(2):687–93.
193. Taylor AH, et al. The cytoplasmic 60 kDa progesterone receptor isoform predominates in the human amniochorion and placenta at term. Reprod Biol Endocrinol. 2009;7:22.
194. Uldbjerg N, Ekman G, Malmstrom A. Ripening of the human uterine cervix related to changes in collagen, glycosaminoglycans and collagenolytic activity. Am J Obstet Gynecol. 1982;147:662.
195. Mochizuki M, Maruo T. Effect of dehydroepiandrosterone sulfate on uterine cervical ripening in late pregnancy. Acta Physiol Hung. 1985;65(3):267–74.
196. Rajabi M, Solomon S, Poole AR. Hormonal regulation of interstitial collagenase in the uterine cervix of the pregnant guinea pig. Endocrinology. 1991;128(2):863–71.
197. Sato T, et al. Hormonal regulation of collagenolysis in uterine cervical fibroblasts. Modulation of synthesis of procollagenase, prostromelysin and tissue inhibitor of metalloproteinases (TIMP) by progesterone and oestradiol-17 beta. Biochem J. 1991;275(Pt 3):645–50.
198. Denison FC, Calder AA, Kelly RW. The action of prostaglandin E2 on the human cervix: stimulation of interleukin 8 and inhibition of secretory leukocyte protease inhibitor. Am J Obstet Gynecol. 1999;180(3 Pt 1):614–20.
199. Carbonne B, et al. Effects of progesterone on prostaglandin E(2)-induced changes in glycosaminoglycan synthesis by human cervical fibroblasts in culture. Mol Hum Reprod. 2000;6(7):661–4.
200. Elovitz MA, Mrinalini C. The use of progestational agents for preterm birth: lessons from a mouse model. Am J Obstet Gynecol. 2006;195(4):1004–10.
201. Facchinetti F, et al. Cervical length changes during preterm cervical ripening: effects of 17-alpha-hydroxyprogesterone caproate. Am J Obstet Gynecol. 2007;196(5):453–4.
202. Chwalisz K, Garfield RE. Regulation of the uterus and cervix during pregnancy and labor. Role of progesterone and nitric oxide. Ann N Y Acad Sci. 1997;828:238–53.
203. Chwalisz K, et al. Cervical ripening in guinea-pigs after a local application of nitric oxide. Hum Reprod. 1997;12(10):2093–101.
204. Chwalisz K, Garfield R. Nitric oxide as the final metabolic mediator of cervical ripening. Hum Reprod. 1998;13:245–52.
205. Word RA, et al. Dynamics of cervical remodeling during pregnancy and parturition: mechanisms and current concepts. Semin Reprod Med. 2007;25(1):69–79.
206. Stiemer B, Elger W. Cervical ripening of the rat in dependence on endocrine milieu; effects of antigestagens. J Perinat Med. 1990;18(6):419–29.
207. Marx SG, et al. Effects of progesterone on iNOS, COX-2, and collagen expression in the cervix. J Histochem Cytochem. 2006;54(6):623–39.
208. Chwalisz K, Garfield R. Antiprogestins in the induction of labor. Ann N Y Acad Sci. 1994;734:387–413.
209. Elliott CL, Brennand JE, Calder AA. The effects of mifepristone on cervical ripening and labor induction in primigravidae. Obstet Gynecol. 1998;92(5):804–9.
210. Stenlund PM, et al. Induction of labor with mifepristone–a randomized, double-blind study versus placebo. Acta Obstet Gynecol Scand. 1999;78(9):793–8.
211. Giacalone PL, et al. The effects of mifepristone on uterine sensitivity to oxytocin and on fetal heart rate patterns. Eur J Obstet Gynecol Reprod Biol. 2001;97(1):30–4.
212. Hegele-Hartung C, et al. Ripening of the uterine cervix of the guinea-pig after treatment with the progesterone antagonist onapristone (ZK 98.299): an electron microscopic study. Hum Reprod. 1989;4(4):369–77.
213. Wolf JP, et al. Progesterone antagonist (RU 486) for cervical dilation, labor induction, and delivery in monkeys: effectiveness in combination with oxytocin. Am J Obstet Gynecol. 1989;160(1):45–7.
214. Stys SJ, Clewell WH, Meschia G. Changes in cervical compliance at parturition independent of uterine activity. Am J Obstet Gynecol. 1978;130(4):414–8.

215. Glassman W, Byam-Smith M, Garfield RE. Changes in rat cervical collagen during gestation and after antiprogesterone treatment as measured in vivo with light-induced autofluorescence. Am J Obstet Gynecol. 1995;173(5):1550–6.
216. Imada K, et al. Hormonal regulation of matrix metalloproteinase 9/gelatinase B gene expression in rabbit uterine cervical fibroblasts. Biol Reprod. 1997;56(3):575–80.
217. Imada K, et al. An antiprogesterone, onapristone, enhances the gene expression of promatrix metalloproteinase 3/prostromelysin-1 in the uterine cervix of pregnant rabbit. Biol Pharm Bull. 2002;25(9):1223–7.
218. Granstrom L, Ekman G, Malmstrom A. Insufficient remodelling of the uterine connective tissue in women with protracted labour. Br J Obstet Gynaecol. 1991;98(12):1212–6.
219. Osmers R, et al. Collagenase activity in the cervix of non-pregnant and pregnant women. Arch Gynecol Obstet. 1990;248(2):75–80.
220. Danforth DN, Buckingham JC, Roddick Jr JW. Connective tissue changes incident to cervical effacement. Am J Obstet Gynecol. 1960;80:939–45.
221. Winn RJ, Baker MD, Sherwood OD. Individual and combined effects of relaxin, estrogen, and progesterone in ovariectomized gilts. I. Effects on the growth, softening, and histological properties of the cervix. Endocrinology. 1994;135(3):1241–9.
222. Clark K, et al. Mifepristone-induced cervical ripening: structural, biomechanical, and molecular events. Am J Obstet Gynecol. 2006;194(5):1391–8.
223. Cabrol D, et al. Prostaglandin E2-induced changes in the distribution of glycosaminoglycans in the isolated rat uterine cervix. Eur J Obstet Gynecol Reprod Biol. 1987;26(4):359–65.
224. Danforth DN, et al. The effect of pregnancy and labor on the human cervix: changes in collagen, glycoproteins, and glycosaminoglycans. Am J Obstet Gynecol. 1974;120(5):641–51.
225. von Maillot K, et al. Changes in the glycosaminoglycans distribution pattern in the human uterine cervix during pregnancy and labor. Am J Obstet Gynecol. 1979;135(4):503–6.
226. Osmers R, et al. Glycosaminoglycans in cervical connective tissue during pregnancy and parturition. Obstet Gynecol. 1993;81(1):88–92.
227. Cabrol D, et al. Induction of labor and cervical maturation using mifepristone (RU 486) in the late pregnant rat. Influence of a cyclooxygenase inhibitor (Diclofenac). Prostaglandins. 1991;42(1):71–9.
228. Rechberger T, Woessner Jr JF. Collagenase, its inhibitors, and decorin in the lower uterine segment in pregnant women. Am J Obstet Gynecol. 1993;168(5):1598–603.
229. Junqueira LC, et al. Morphologic and histochemical evidence for the occurrence of collagenolysis and for the role of neutrophilic polymorphonuclear leukocytes during cervical dilation. Am J Obstet Gynecol. 1980;138(3):273–81.
230. Hertelendy F, Zakar T. Prostaglandins and the myometrium and cervix. Prostaglandins Leukot Essent Fatty Acids. 2004;70(2):207–22.
231. Ramos JG, et al. Estrogen and progesterone modulation of eosinophilic infiltration of the rat uterine cervix. Steroids. 2000;65(7):409–14.
232. Barclay CG, et al. Interleukin-8 production by the human cervix. Am J Obstet Gynecol. 1993;169(3):625–32.
233. Baggiolini M, Walz A, Kunkel SL. Neutrophil-activating peptide-1/interleukin 8, a novel cytokine that activates neutrophils. J Clin Invest. 1989;84(4):1045–9.
234. Leonhardt SA, Boonyaratanakornkit V, Edwards DP. Progesterone receptor transcription and non-transcription signaling mechanisms. Steroids. 2003;68(10–13):761–70.
235. Yen SSC. Endocrine-metabolic adaptation in pregnancy. In: Yen SSC, Jaffe RB, editors. Reproductive endocrinology. Philadelphia, PA: WB Saunders; 1991. p. 936–71.

Chapter 2
Pharmacology of Progestogens

Adolf E. Schindler

1 Introduction

Progesterone is the only natural progestogen, synthesized, produced and released by the corpus luteum of the ovary during the luteal phase. Its unique features include, an increase in the basal body temperature, secretory changes in the endometrium in preparation for fertilisation and ovum implantation. In addition to these basic biological effects, the progestogens may, to differing degrees, suppress the hypothalamic pituitary axis, an effect which mainly accounts for the contraceptive effect of some progestogens. Progestogens may also affect abnormal endometrial tissue such as hyperplastic endometrium and endometriosis which have been chronically and excessively stimulated by endogenous or exogenous estrogens.

In 1934 progesterone was isolated from animal corpora lutea and structurally identified [1]. The structure of progesterone is shown in Fig. 2.1. In the 1940s the manufacture of progesterone was made possible, by synthesis from the plant sterol "diosgenin" [1]. However, it was soon realized that progesterone could not be properly absorbed by the gut. Progress in pharmacological development led to the development of "micronization" of progesterone, which improves oral as well as parenteral absorption [2]. However, bioavailability is low (approximately 5 %). Other progestogens, have different absorbative properties and different bioavailabilities. Dydrogesterone for example, a retroprogesterone, derived from progesterone by ultra violet light exposure is readily absorbed orally and has been calculated to have 28 % bioavailability.

The broad variety of progestogens available with different progestogenic potencies, as well as a whole array of partial effects, warrants a detailed presentation and description of the progestogens.

A.E. Schindler, M.D., Ph.D., M.D.(h.c.) (✉)
Institute for Medical Research and Education, University Clinic Essen, SHH1,
Hufelandstrasse 55, D-45122 Essen, Germany
e-mail: adolf.schindler@uni-due.de

H.J.A. Carp (ed.), *Progestogens in Obstetrics and Gynecology*,
DOI 10.1007/978-3-319-14385-9_2

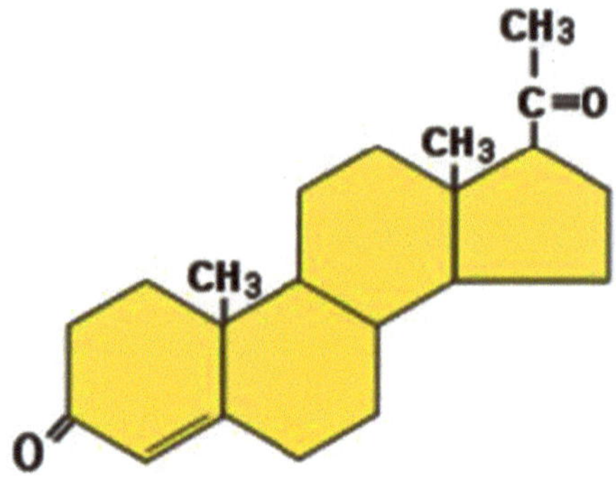

Fig. 2.1 Structure of progesterone

Table 2.1 Progestogens as prodrugs

Prodrug	Clinically relevant compound
Norethinodrel	Norethisterone
Trimegestone	Promegestone
Tibolone	3αOH-tibolone 3βOH-tibolone Δ4-isomer
Desogestrel	3-Keto-desogestrel
Norgestimate	Levonorgestrel

2 Classification of Progestogens

Progestogens are all steroid compounds with progestogenic activity of different intensity regarding progestogenic action per molecule. Some of the compounds used therapeutically act as prodrugs, which need to be metabolized in the body before the respective function is obtained (see Table 2.1). Progestogens not only have different structures, but each progestogen expresses a different pattern of partial effects. This partial effect pattern is responsible for the different clinical effects and side effects seen for each progestogen.

Progesterone and all the other synthetic progestogens are called progestogens. If progesterone is excluded the term "progestin" is used. An overview of the various types of progestogens is shown in Table 2.2 [3, 4]. The development of the classification of the progestogens started in the 1950s. Removal of the C19-Methyl-group increased progestogenic activity and oral resorption but decreased androgenic action. The introduction of a 17α-Ethyl-group produced ethisterone, which had a much higher binding affinity to the progesterone receptor. Both processes together produced the progestogen norethisterone (norethindrone in the United States), which is highly active, well tolerated and has been clinically available since 1957 [5]. In 1951 Norethisterone acetate (NETA) was synthesized by Schering and Norethinodrel by Searle. In the 1960s, the prodrugs norethisterone (NET), lynestrenol, ethinylethinodiol acetate, norethinodrel and DL-norgestrel appeared on the market.

Table 2.2 Classification of progestogens, and partial effect pattern modified from [3, 4]

Progestogen	Anti-gonadotrophic	Estrogenic		Androgenic		Glucocorticoid	Anti mineralocorticoid
		Pro	Anti	Pro	Anti		
Progesterone	+	–	+	–	+	±	±
Pregnane derivatives: non acetylated							
Dydrogestrone	–	–	+	–	±	–	±
Medrogesterone	+	–	+	–	±		
Pregnane derivatives: acetylated							
Medroxyprogesterone acetate	+	–	+	±	–	+	–
Megestrol acetate	+	–	+	±	+	+	–
Chlormadinone acetate	+	–	+	–	+	+	–
Cyproterone acetate	+	–	+	–	++	+	–
19-Norpregnane derivatives: non acetylated							
Demegestone	+	–	+	–	–	–	–
Promegestone	+	–	+	–	–	–	–
Trimegestone	+	–	+	–	±	–	±
19-Norpregnane derivatives: acetylated							
Nomegestrol acetate	+	–	+	+	±	–	–
Nesterone	+	–	+	–	–	–	–
19-Nortestosterone derivatives: Estranes							
Norethisterone (Norethindrone)	+	+	+	+	–	–	–
Norethisterone acetate	+	+	+	+	–	–	–
Norethynodrel	+	+	±	±	–	–	–
Lynestrenol	+	+	+	+	–	–	–
Tibolone (metabolites)	+	+	–	+	–	–	–
Dienogest	+	±	+	–	+	–	–

(continued)

Table 2.2 (continued)

Progestogen	Anti-gonadotrophic	Estrogenic		Androgenic		Glucocorticoid	Anti mineralocorticoid
		Pro	Anti	Pro	Anti		
19-Nortestosterone derivatives: Gonanes							
Levonorgestrel	+	–	+	+	–	–	–
Norgestimate	+	–	+	+	–	–	–
Desogestrel (etogestrel)	+	–	+	+	–	–	–
Gestodene	+	–	+	+	–	+	+
Spirolactone derivative							
Drospirenone	+	–	+	–	+	–	+

++, strongly positive; +, positive; ±, weakly positive; –, negative

The first progesterone derivative 17-acetoxyprogesterone was developed by Schering in 1954 followed by medroxyprogesterone acetate in 1957. This was followed by medrogestone acetate and chlormadinone acetate in 1959. The retroprogesterone dydrogesterone was formed from progesterone by UV light exposure [6]. Dydrogesterone is still the only retroprogesterone presently available on the market. In 1961 cyproterone acetate became clinically available, followed by desogestrel in 1972 (Organon). Thereafter gestodene, dienogest and the spirolactone derivative drospirenone followed [7].

3 Pharmacokinetics and Pharmacology of Progestogens

Pharmacokinetics such as absorption, distribution and excretion determine how much of the progestogen is available to the tissues, by measuring the blood levels and the amount that enters the cell is regulated by the extent to which the progestogen is bound to carrier proteins. Carrier proteins cannot cross the cell membranes. The pattern of distribution of the progestogens is mainly regulated by binding to transport proteins and steroid receptors in the tissues.

Generally, all progestogens are bound in the blood with low affinity and high capacity to albumin. However, some of the progestogens derived from 19-Nortestosterone, such as norethisterone (norethindrone) are also bound with high affinity but low capacity to sex hormone binding globulin (SHBG), while others, such as progesterone itself can be bound to the corticosteroid-binding globulin (CBG). The binding of progestogens to transport proteins is reversible, so that a change in binding protein concentration may contribute to the variation or variability of a progestogen. The non-protein-bound (unbound or free fraction) of a steroid is available for metabolism in steroid metabolising cells or binds to a receptor in target cells.

Progestogens given orally reach a maximum concentration within 1–3 h. Information on bioavailability and half-life has been derived from frequent blood sampling during the first 24 h after oral administration. Bioavailability represents the amount of the progestogen that is found in the circulation (area under the curve). The half-life is the time in hours in which the progestogen has been absorbed to one half of its highest level. The longest half life is found with drospirenone (31–32 h), whereas norethisterone has the shortest half life (8 h). Details are summarized in Fig. 2.2.

Among progestogens there are great differences in bioavailability. Progesterone itself has a bioavailability of less than 5 %, dydrogestrone has a bioavailability of 28 % and nomegestrol of 60 % [4]. The bioavailability of progestogens derived from 19-nortestosterone can reach more than 90 %. The distribution of some progestogens bound to SHBG, CBG, albumin and free fraction is shown in Table 2.3.

The clinical effects of the progestogens is not only dose dependent but also influenced by the different partial effect pattern of each progestogen, as summarized in Table 2.2. Each progestogen has a different partial effect pattern, which can modify the final biological effect of each progestogen. Acquaintance with the partial effect pattern will enable the clinician to choose the optimal progestogen.

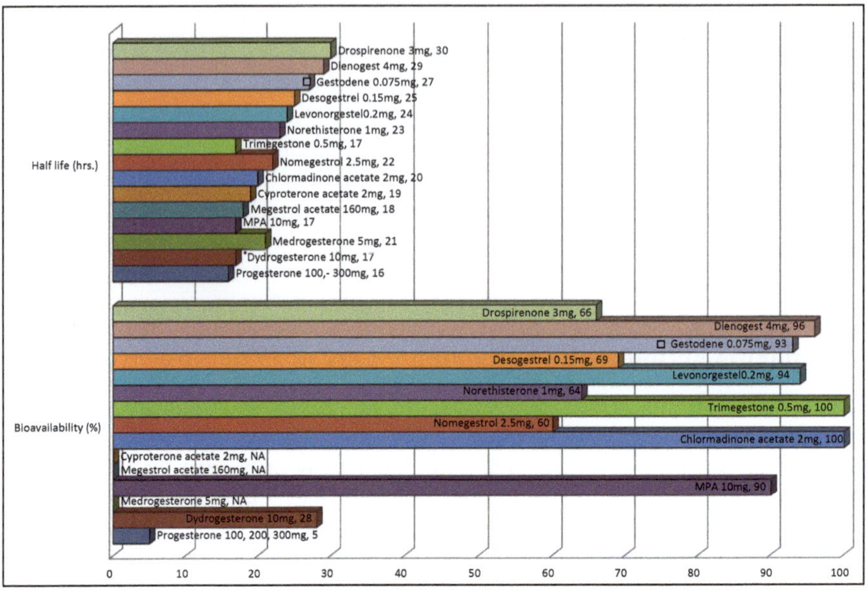

Fig. 2.2 Dose, bioavailability and half-life of progestogens modified from [3, 4]. *Dydrogesterone 17 h with metabolites

Table 2.3 Relative binding affinities of progestogens to steroid receptors and serum binding proteins modified from [3, 7]

	PR	AR	ER	GR	MR	SHBG	CBG	Albumin bound	Free
Progesterone	50	0	0	10	100	0	36	79.3	2.4
Dydrogesterone	75	0	–	–	–	–	–		
Chlormadinone acetate	67	5	0	8	0	0	0		
Cyproterone acetate	90	6	0	6	8	0	0		
Medroxyprogesterone acetate	115	5	0	29	160	0	0		
Megestrol acetate	65	5	0	30	0	0	0		
Nomegestrol	125	6	0	6	0	0	0		
Promegestone (R5020)	100	0	0	5	53	0	0		
Drospirenone	35	65	0	6	230	0	0		
Norethisterone	75	15	0	0	0	16	0	60.8	3.7
Levonorgestrel	150	45	0	1	75	50	0	50	2.5
Norgestimate	15	0	0	1	0	0	0		
Desogestrel (Etonogestrel)	150	20	0	14	0	15	0	65.5	2.5
Gestodene	90	85	0	27	290	40	0	24.1	0.6
Dienogest	5	10	0	1	0	0	0		

PR progesterone receptor (promegestone = 100 %), *AR* androgen receptor (metribolone = 100 %), *ER* estrogen receptor (estradiol-17β = 100 %), *GR* glucocorticoid receptor (dexamethason = 100 %), *MR* mineralocorticoid receptor (aldosterone = 100 %), *SHBG* sex hormone binding globulin (dihydrotestosterone = 100 %), *CBG* corticosteroid-binding globulin (cortisol = 100 %), *ND* not determined

Table 2.4 Comparison of partial effects and metabolic effects of dydrogesterone, medroxyprogesterone acetate and norethisterone

Progestogen	Dydrogesterone	MPA	Norethisterone (Norethindrone)
Androgenic	No	Mildly	Yes
Estrogenic	No	No	Metabolites
Glucocorticoid	No	Yes	No
HDL cholesterol	No effect	↓ (reduces E effect)	↓↓ (androgen effect)
Glucose metabolism	No effect	↓ glucose tolerance	↓ glucose tolerance

Progestogens also differ according to the affinity for various steroid receptors such as the progesterone receptor (PR), estrogen receptor (ER), androgen receptor (AR), mineralocorticoid receptor (MR) or glucocorticoid receptor (GR). Affinity for the different receptors is summarised in Table 2.3. Affinity for different receptors is influenced by specific receptor binding proteins (Table 2.3). The clinical consequences of receptor binding are shown in Table 2.4.

A most recent area of interest and controversies is the influence of the various progestogens on thromboembolic risk [8].

4 Conclusions

Progestogens are different: in structure and in action profile. Besides the common progestogenic effect, each progestogen has a particular partial effect pattern, which has utmost relevance when clinically used. Effects and possible side effects can be influenced or determined by this.

References

1. Williams CL, Stancel GM. Estrogens and progestins. In: Goodman L, Gilman A, editors. The pharmacological basis of therapeutics. Elmsford, Oxford: Pergamon Press; 1996. p. 1411–40.
2. Morville R, Dray F, Regnier J, Barrat J. The bioavailability of natural progesterone given by month. Measurements of steroid concentration in plasma, endometrium and breast tissue. J Gynecol Obstet Biol Reprod (Paris). 1982;11:355–63.
3. Schindler AE, Campagnoli C, Druckmann R, et al. Classification and pharmacology of progestogens. Maturitas. 2003;46 Suppl 1:S7–16.
4. Stanczyk FZ, Hapgood JP, Winter SH, et al. Progestogens used in postmenopausal hormone therapy: differences in their pharmacological properties, intracellular actions and clinical effects. Endocr Rev. 2013;34:171–208.
5. Kuhl H. Pharmacology of progestogens. J Reproduktionsmed Endokrinol. 2011;8 Suppl 1:157–76.

6. Reesink EH, Schöler HFL, Westerhof P, et al. A new class of hormonally active steroids. Nature. 1960;186:168–9.
7. Wiechart R. Analogue based drug discovery. In: Fisher J, Robin Ganellin C, editors. IUPAC. Weinheim: Wiley VCH; 2006. p. 395–400.
8. Schindler AE. Progestogens and thromboembolic risk. Front Gynecol Endocrinol. 3: 2014 (in press)

Chapter 3
Progestogens in Infertility Practice

Ameet S. Patki and Alok Sharma

1 Introduction

The physiological role of progesterone is to prepare the endometrium for implantation and to support pregnancy. In fact, the name progesterone is derived from the Latin word 'Gestare' meaning to bear or carry. It is also believed that the name progesterone is derived from progestational steroidal ketone [1]. Progesterone is secreted primarily from the corpus luteum of the ovary during the second half of the menstrual cycle and from the placenta during pregnancy. Progesterone's role in mammalian reproduction is well described and undisputed. After ovulation, progesterone, secreted by the corpus luteum induces transformation of the proliferative endometrium into the secretory type required for implantation. The endometrium then undergoes specific morphological changes, which are termed 'decidualization'. All cellular types and structures localized in the functional layer are targets for progesterone action, namely, stromal cells, epithelial glands, and spiral arteries. However, in addition to the endocrine effects, progesterone has numerous immuno-modulatory effects.

In 1976 Georgeanmna Seegar Jones first described luteal deficiency (also known as luteal phase defect or luteal insufficiency), as a condition in which the corpus luteum produces inadequate amounts of progesterone for implantation, or placentation, or a

A.S. Patki, M.D., D.N.B., F.C.P.S., F.R.C.O.G. (✉)
Fertility Associates, Mumbai, India

K.J. Somaiya Medical College and Hospital, Mumbai, India

Hinduja Surgical Hospitals, Mumbai, India

The Mumbai Obstetrics and Gynaecological Society (2014–15), Mumbai, India
e-mail: aspaap@hotmail.com; ameetpatki@fertilityassociates.in

A. Sharma, M.D.
Clinical Fellow Fertility Associates, Mumbai, India

H.J.A. Carp (ed.), *Progestogens in Obstetrics and Gynecology*,
DOI 10.1007/978-3-319-14385-9_3

lack of an adequate endometrial response [2], due to a suboptimal number of receptors. In these conditions progesterone or progestogen supplementation has been used. Luteal phase deficiency has been claimed to be responsible for: subfertility, implantation failure, and recurrent miscarriage. However, the whole concept of luteal deficiency has been controversial since its inception, with many doubting the existence of the condition. Similarly progesterone supplementation has also been controversial, with numerous workers doubting any beneficial effects. This chapter will examine the role of luteal deficiency in clinical infertility practice in general, and assisted reproduction in particular.

2 Role of Progesterone in Endometrial Ripening

In the proliferative phase of the cycle, estrogen induces endometrial proliferation. In the luteal phase, progesterone induces changes in endometrial morphology converting the proliferative endometrium to a secretory endometrium. In the secretory phase, the endometrial glands and blood vessels become more tortuous. Glycogen accumulates in vacuoles within the glandular cells, leading to secretion of glycoproteins and peptides into the endometrial cavity, and the stroma becomes edematous. Under the influence of progesterone, stromal cells are transformed into decidual cells, with accompanying infiltration of natural killer (NK) cells, T cells, and macrophages. Pinopode formation coincides with increased progesterone levels and down-regulation of progesterone receptors during the window of implantation Pinopodes may extract fluid from the uterus, facilitating closer contact between blastocyst and endometrium. Progesterone increases osteopontin (OPN, a ligand for integrin $\alpha_v\beta_3$ secretion) a bridging molecule between the embryo and endometrium. In the mid luteal phase, Leukemia inhibitory factor (LIF, a cytokine which is essential for implantation in muridae) is upregulated. In fact, antiprogestin treatment results in reduced LIF expression. HOXA-10 and 11 genes are up-regulated by estrogen and progesterone. HOXA-10 mediates integrin involvement in early embryo–endometrial interactions. HOXA-10 expression is required for pinopode formation in the mouse.

3 Role of Progesterone in Implantation

Normal luteal function is essential for initiating pregnancy. After adequate estrogen priming, progesterone induces secretory transformation of the endometrium which improves endometrial receptivity [3, 4]. In the "window of implantation" the endometrial epithelium acquires a functional and transient ovarian steroid-dependent status, which allows blastocyst adhesion [5]. The complex molecular interactions between the hormonally-primed uterus and an activated blastocyst results in successful implantation. Progesterone induces the proliferation and differentiation of stromal cells [6]. Progesterone receptor synthesis is controlled by estrogens through estrogen receptors during the proliferative phase. If the synthesis of the estrogen receptors is inhibited then progesterone leads to a fall of both estrogen and progesterone receptors. Various experimental studies have reported down regulation of progesterone receptor epithelial cell expression during the luteal phase of the menstrual cycle [7].

Local vasodilatation and uterine musculature quiescence is also promoted by progesterone by inducing nitric oxide synthesis in the decidua [8]. Fanchin et al. [9] investigated the consequences of uterine contractions at the time of embryo transfer. The authors reported that on the day of embryo transfer, a high frequency of uterine contractions hindered the transfer outcome, possibly by expelling the embryos from the uterine cavity. Additionally, there was a negative correlation between the frequency of contractions and progesterone concentrations [3].

Decreased endometrial receptivity is the main factor responsible for the low implantation rates in IVF [10].

3.1 Cytokines Acting in Implantation

Numerous cytokines are active in implantation, in order to modulate the inflammatory response, remodel tissues, and to induce endocrine effects. The entire picture is far from complete, but some of the cytokine effects are listed below. Cytokines such as Interleukin (IL)-3, Granulocyte Macrophage Colony Stimulating Factor (GM-CSF) and Epidermal Growth Factor (EGF) stimulate placental cell proliferation [11] in vitro, and may enable the trophoblast to secrete hCG and hPL [12]. Interferon γ (IFNγ) leads to remodelling of the spiral arteries to utero-placental arteriies [13]. Interleukin-1 (IL-1) has many effects, both proinflammatory and anti-inflammatory. IL-1 stimulates IL-6, IL-8, LIF, TNFα, PGE2, PGF production. IL-1 induce COX-2 gene expression which mediates prostaglandin synthesis, induces MMP-1 production and increases the activity of MMP-9, which are involved in implantation. IL-1 modulates hCG and CRH synthesis and attenuates progesterone production by granulosa cells Additionally in animal models, blocking IL-1 reduced the number of implanted blastocysts. Interleukin-6 (IL-6) is associated with hCG release, and increases MMP9.

Leukemia inhibitory factor (LIF) is essential for implantation. LIF mRNA is induced by estrogen. Female mice with no LIF gene, have normal blastocysts which fail to implant. Injecting $LIF^{-/-}$ mice with LIF causes viable pregnancies. In humans, LIF causes cytotrophoblast differentiation to an anchoring phenotype due to increased fibronectin synthesis. Infertile patients with multiple implantation failure have been reported to have dysfunctional LIF production compared to fertile women [14]. Interleukin-15 (IL-15) increases trophoblast invasion, modulates MMP-1 and maintains uterine natural killer (NK) cells.

3.2 Action of Progesterone on Cytokines

Progesterone has been shown to induce changes in the functions of a number of immune-competent cells by different molecular and cellular mechanisms. Progesterone stimulates the activity of some specific enzyme matrix metalloproteinases [15]. and adhesion molecules [16], inhibits antibody production and suppresses T-cell activation and cytotoxicity [17] and directly or indirectly modifies the activity of NK cells, which are the most numerous lymphoid cells locally [18].

Progesterone is associated with decreased IFNγ and increased IL-10 in endocervical fluid [19]. Progesterone up regulates LIF mRNA expression in vitro [20]. Progesterone inhibits NK activity at the feto-maternal interface, inhibits the release of arachidonic acid, and favours the production of asymmetric, pregnancy-protecting antibodies. The cytokine effects of the progestogen dydrogesterone have been investigated more than progesterone itself. Dydrogesterone, inhibits IFNγ and TNFα production [21]. Dydrogesterone increases the levels of IL-4 and IL-6 [21]. Dydrogesterone Inhibits NK activity at the feto-maternal interface, and in preterm labour, dydrogesterone is associated with significantly higher serum levels of IL-10, and lower concentrations of IFNγ than controls [22].

4 Luteal Phase Insufficiency

4.1 Formation of the Corpus Luteum

The CL contains a heterogeneous population of cells including steroidogenic cells, fibroblasts, immune cells, and endothelial cells. The steroidogenic cells are the large and the small luteal cells that are derived from the granulosa and theca cells of the ruptured follicle, respectively [23]. The granulosa cells of the pre-ovulatory follicle are not vascularized, the blood supply stops at the basement membrane. Following ovulation, basement membrane integrity is lost, tissue re-modeling takes place and vessels, originating from the thecal vasculature, invade the granulosa-luteal cells [24]. Over the next few days, intensive angiogenesis takes place and a capillary network extends throughout the fully differentiated CL tissue. In humans, both vascular density and endothelial area of each vessel increase markedly from the luteinized granulosa cells of the early CL to the mid-luteal stage [25]. Neoangiogenesis is important for CL function and is controlled by various angiogenic factors, such as vascular endothelial growth (VEGF), fibroblast growth factor, angiopoietins, and insulin-like growth factors.

The process of neovascularization is regulated by pituitary LH [26]. Luteinizing hormone activates matrix metalloproteinases that degrade extracellular matrix associated with the blood vessels. The CL produces progesterone, estrogens, and non-steroidal substances, such as inhibin A. Apart from the pituitary gonadotropins, different local substances may also regulate the CL life span and function. Such substances include growth factors, peptides, steroids, and prostaglandins [26].

Normal luteal function requires optimal pre-ovulatory follicular development, luteinization of the granulosa cells to produce progesterone, continued tonic luteinizing hormone (LH) support, vascularization of the corpus luteum (CL), and estrogen to induce progesterone (P4) receptors in the endometrium [27]. The normal luteal life span is 14 days unless the life span is prolonged by human chorionic gonadotropin (hCG). hCG is secreted by the developing blastocyst after implantation [28]. By inducing secretory changes, following adequate estrogen priming, progesterone induces normal endometrial receptivity.

4.2 Evaluation of the Luteal Phase

Luteal phase insufficiency generally stems from an insufficiency of estrogen and progesterone production after ovulation. At mid-cycle, the gonadotropin surge has important physiological roles, including induction of luteinization of the granulosa cells, resumption of oocyte meiosis, rupture of the pre-ovulatory follicle, and formation of the CL. Among other events, post-LH surge changes include a shift in steroidogenesis within the follicle with a marked decrease in estradiol (E2) concentrations and a gradual increase in serum progesterone concentrations [29, 30]. Additional alterations involve uncoupling of gap junctions between granulosa cells and the plasma membrane of the oocyte, a process that seems to be important for the resumption of meiosis [31]. For the transfer of cholesterol from the outer to the inner surface of the mitochondrial membrane steroidogenic acute regulatory protein (StAR) is important [32]. The StAR protein is absent from the granulosa cells before the onset of the LH surge and this explains the inability of these cells to produce progesterone [33].

Various investigations have been suggested to assess luteal insufficiency. The original method was by endometrial biopsy taken 2 days prior to menstruation and histologically dated according to Noyes et al.'s [34] criteria. Luteal deficiency is thought to occur if the dating lags more than 2 days behind the chronological age. This technique has produced a positive diagnosis in 26–35 % of cases [35, 36]. However, if the dating is accurately determined by ultrasonic monitoring of follicular size and biopsy timed to 12 days later, the incidence is only 4 % [37]. However, biopsy is taken in a non pregnancy cycle and assumed to reflect the situation in a pregnancy cycle. Plasma progesterone levels are an unreliable test of luteal function as progesterone is secreted in a pulsatile fashion, and blood may be drawn at a pulse peak or nadir. These may differ tenfold [38]. There may also be normal hormone levels in the presence of abnormal histology may also be due to a deficiency of progesterone receptors rather than a deficiency in progesterone itself.

4.3 Luteal Insufficiency in Stimulated Cycles

Luteal insufficiency is due to the inhibition of the LH secretion in the early luteal phase by steroids secreted in supra-physiologic doses in stimulated cycles [39]. If luteal phase hormonal support is not present in assisted reproduction technique (ART) cycles, the serum estrogen and P4 levels drop, thus leading to a decrease in the implantation rates and pregnancy rates [40]. There may be two types of luteal phase defect: one is associated with the presence of immature follicles, and the other where the follicles are mature. In both types, supplemental therapy with progesterone is effective in creating a healthy uterine environment [41].

In women undergoing ovulation induction, multiple follicles of different size might ovulate at different times, thus expanding the fertilization window. It can be expected that sex steroid concentrations, both estradiol (E2 and progesterone), after multiple ovulation will be significantly higher [42]. These high concentrations may not only influence the receptivity of the endometrium, but may also cause luteal insufficiency [43] as high concentrations of steroids through negative feedback on the pituitary-hypothalamic axis, inhibit the production of luteal LH, which is mandatory for luteal progesterone production. A significant negative correlation has been reported between both pre-ovulatory estradiol concentrations and day 16 progesterone levels and the concentration of cytosolic progesterone receptor (Cpr) [44], while advanced endometrial maturity tends to be associated with low concentrations of cPR. Furthermore, natural cycles have been characterized by low cytosolic E-2 receptors (cER) and high cPR, whereas the concentration of both receptors was greatly reduced in stimulated cycles. Due to receptor abnormalities, the endometrium can be progesterone deficient even if plasma progesterone levels are normal [45]. According to one study it has been suggested that in approximately 25 % of women with recurrent miscarriage, the spatial expression of progesterone receptors in the endometrium is different from that in normal controls. Impaired reproductive function and early abortions may be caused by mutations in the progesterone receptor gene [46]. Polymorphism within the coding sequence of the human progesterone receptor gene hs been reported to be significantly higher in patients with recurrent miscarriages than in controls.

5 Hormone Supplementation in Infertility

In normal ovulatory sub-fertile women, with primary or secondary infertility, 92 % of cycles show normal luteal function. Therefore, luteal support seems to be unnecessary [47]. However, latrogenic LPD is seen with the use of controlled ovarian stimulation, and gonadotropin-releasing hormone (GnRH) analogues for in vitro fertilization (IVF), and intrauterine insemination (IUI) cycles. Latrogenic LPD has provided an opportunity to study the endocrine and endometrial abnormalities during the luteal phase and the impact of any pharmacological intervention. Various regimens of hormone supplementation have been used. These are discussed below.

Table 3.1 Progestogens in luteal phase support

	OR	Side effects	Studies (women)
hCG vs. Placebo or No Rx	1.30 (0.09–1.88)	OHSS	5 (746)
Progestin/placebo, No Rx	1.83 (1.29–2.61)		7 (741)
Progestin vs. hCG	1.14 (0.90–1.45)	OHSS	10 (2,117)
Progestin/progestin, GnRHa	1.36 (1.11–1.66)		6 (1,646)
Micronised vs. DYD	0.79 (0.65–0.96)		32 (9,839)

Adapted from Van der Linden et al. [51]
Results refer to clinical pregnancy rates

5.1 *Human Chorionic Gonadotropin (hCG)*

Of the various mechanisms by which hCG may rescue the corpus luteum, an increase in both E_2 and progesterone levels appears to be the most likely [48]. The usual dose is 1,500–2,500 IU twice a week from the day of embryo transfer (ET) continued until the day of the pregnancy test or until 8–10 weeks of gestation. Meta-analyses comparing hCG with progesterone have shown it to be associated with either better or at least similar pregnancy rates to that seen with progesterone [49, 50]. Recently, Van der Linden et al. [51] published the results of a metaanalysis to assess luteal phase support (progestogen, hCG or GnRH agonists) in the Cochrane database. The review comprised 69 studies of 16,327 women. The results are summarised in Table 3.1. hCG administration did not improve the pregnancy rate significantly when compared to placebo (OR = 1.30 95 % CI 0.09–1.88). and showed similar results to progesterone supplementation [51]. However, the risk of ovarian hyper-stimulation syndrome (OHSS) associated with hCG in stimulated IVF cycles limits its use for luteal support [52]. The results of luteal phase dynamics differ after GnRHa trigger in GnRH antagonist-treated cycles, the luteal phase being short or inadequate. Whether hCG offers a safe and effective luteal support in this group of women without the risk of OHSS is yet to be fully understood [53].

5.2 *Micronized Progesterone*

Micronized progesterone is to-day the most widely used form of luteal support. Micronized progesterone can be administered orally, rectally or vaginally. However, the bioavailability of micronized progesterone following oral administration is variable as progesterone is metabolised in the liver to pregnenolone and pregnanediol and thereby inactivated. Hence, endometrial changes are inconsistent [54]. In addition, side effects such as nausea, abdominal bloatedness, drowsiness are common with oral administration. The rectal route is rarely used. The vaginal route of administration is widely used, due to ease of administration and high bioavailability as hepatic degradation is avoided. Intra-vaginal administration results in a high uterine concentration of progesterone with relatively low levels in the peripheral circulation. Vaginal micronised progesterone is available in both capsule and gel forms. The daily dose is 600–800 mg/day in 2–3 divided doses, although no dosage finding study has been performed; and 90 mg of gel (8 %) once daily. Pregnancy rates are similar with both forms of vaginal preparations [55, 56]. The disadvantages of vaginal micronized progesterone include local irritation in some women, discharge from the gel or capsule, and staining of the clothes. The divided dose may also be inconvenient, as daily routines need to be interrupted in the middle of the day. Additionally, vaginal administration is unacceptable in some societies.

When all progestogens were compared to placebo or no treatment in Van der Linden et al.'s [51] metaanalysis, progestogens were found to give a significantly better pregnancy rate than placebo or no treatment (OR = 1.83 95 % CI 1.29–2.61).

5.3 *Intramuscular Progesterone*

Progesterone in oil, 50–100 mg daily as an intramuscular injection is another form of luteal support. With the availability of vaginal progesterone, the intramuscular route is less often used than previously. Pain, rash and abscess at the injection site and the need for daily visits for intramuscular injection by trained staff, are important factors precluding routine use. Occasional occurrence of eosinophilic pneumonia has been reported in otherwise healthy women. However, if the vaginal route of administration is unacceptable or if there is severe local irritation, the i.m. route of administration may be preferred to the vaginal route. More recent evidence has found both routes to be equally effective [51, 57].

5.4 *Estradiol Plus Progesterone*

Luteal estradiol supplementation has been used in addition to progesterone support in an attempt to improve IVF outcomes, both in women with low luteal estrogen levels or electively in all treatment cycles. Transdermal estradiol patches delivering a dose of 100 μg/day or oral or vaginal estradiol 4–6 mg/day together with progesterone have all been used with variable results. The current evidence of benefit is limited to a higher implantation rate seen in one single study [58]. The addition of estradiol to progesterone support has not been shown to improve the pregnancy rate [51, 59].

5.5 *Progesterone with Gonadotrophin-Releasing Hormone*

Small bolus doses of GnRH have been used in an attempt to improve pregnancy rates in antagonist-treated cycles where GnRHa is used to trigger ovulation. Triptorelin 0.1 mg has been administered on the day of oocyte pick-up, embryo transfer and 3 days afterwards. The original reports suggested an improvement in both pregnancy and live birth rates [51, 60, 61]. GnRHa may support the CL by stimulating the secretion of LH by pituitary gonadotroph cells by acting directly on the endometrium through locally expressed GnRH receptors. Luteal-Phase GnRHa administration increases luteal phase serum hGC, estradiol and progesterone concentrations. The beneficial effect could possibly be due to a combination of effects on the embryo and the CL.

5.6 Synthetic Progestogens

Synthetic progestogens derived from 19-nor testosterone have stimulatory effects on the androgen receptors. Therefore, although effective, these preparations are not recommended in infertility practice for the fear of inducing androgenic side effects on a female fetus. Androgenization of a female embryo has been seen in laboratory rats, but has never been reported in humans.

Dydrogesterone is a stereoisomer of progesterone, manufactured by conversion of progesterone with ultra violet light. It has been extensively used in over 90 million women, in 90 countries over 40 years and has been found safe for use in pregnancy [62]. Dydrogesterone has a 50 % higher affinity for the progesterone receptor than progesterone itself [63], and has no stimulatory or inhibitory effect on the androgen receptor. Chakravarty et al. [64], have compared oral dydrogesterone to micronised progesterone and found them to be equally effective. Two studies have reported dydrogesterone to have a superior effect to progesterone itself. Iwase et al. [65], has found dydrogesterone to be associated with a significantly higher Live Birth Rate. Patki and Pawar [66] have found a statistically significant increase in pregnancy rates with use of 30 mg dydrogesterone as compared to vaginal micronised progesterone in ART cycles. The Cochrane review of Van der Linden et al. [51] also found a significant effect in favour of dydrogesterone over progesterone itself in terms of pregnancy rates.

Dydrogesterone also has other advantages. It is available as an oral preparation. Although metabolised in the liver, the metabolite dihydrodydrogesterone, is active on the progesterone receptor, unlike the metabolites of progesterone itself.

References

1. Allen WM. The isolation of crystalline progestin. Science. 1935;82:89–93.
2. Jones GES. The luteal phase defect. Fertil Steril. 1976;27:351–6.
3. Kolibianakis EM, Devroey P. The luteal phase after ovarian stimulation. Reprod Biomed Online. 2002;5 Suppl 1:26–35.
4. Kolibianks EM, Devroey P. Blastocyst culture: facts and fiction. Reprod Biomed Online. 2002;5:285–93.
5. Martin' J, Dominguez F, Avila S, et al. Human endometrial receptivity gene regulation. J Reprod Immunol. 2002;55:131–9.

6. Norwitz ER, Schust DJ, Fisher SJ. Implantation and the survival of early pregnancy. N Engl J Med. 2001;345:1400–8.
7. Tuckerman E, Laird SM, Steward R, et al. Markers of endometrial function in women with unexplained recurrent pregnancy loss: a comparison between morphologically normal and retarded endometrium. Hum Reprod. 2004;19:196–205.
8. Bulletti C, de Ziegler D. Uterine contractility and embryo implantation. Curr Opin Obstet Gynecol. 2005;7:265–76.
9. Fanchin R, Righini C, Olivennes F, et al. Uterine contractions at the time of embryo transfer alter pregnancy rates after in-vitro fertilization. Hum Reprod. 1998;13:1968–74.
10. Paulson RJ, Sauer MV, Lobo RA. Embryo implantation after human in vitro fertilization: importance of endometrial receptivity. Fertil Steril. 1990;53:870–4.
11. Chaouat G, Menu E, Wegmann TG. Role of lymphokines of the CSF family and of TNF, gamma interferon and IL-2 in placental growth and fetal survival studied in two murine models of spontaneous resorptions. In: Chaouat G, Mowbray JF, editors. Cellular and molecular biology of the maternal-fetal relationship. Paris: INSERM/John Libbey Eurotext; 1991. p. 91.
12. Garcia-Lloret MI, Morrish DW, Wegmann TG, Honore L, Turner AR, Guilbert LJ. Demonstration of functional cytokine-placental interactions: CSF-1 and GM-CSF stimulate human cytotrophoblast differentiation and peptide hormone secretion. Exp Cell Res. 1994;214:46–54.
13. Ashkar AA, Di Santo JP, Croy AB. Interferon γ contributes to initiation of uterine vascular modification, decidual integrity, and uterine natural killer cell maturation during normal murine pregnancy. J Exp Med. 2000;192:259–70.
14. Hambartsoumian E. Endometrial leukemia inhibitory factor (LIF) as a possible cause of unexplained infertility and multiple failures of implantation. Am J Reprod Immunol. 1998; 39:137–43.
15. Curry TE, Osteen KG. The matrix metalloproteinase system: changes, regulation, and impact throughout the ovarian and uterine reproductive cycle. Endocr Rev. 2003;24:428–65.
16. Aplin JD. Adhesion molecules in implantation. Rev Reprod. 1997;2:84–112.
17. Lin H, Mosmannn TR, Guilbert L, et al. Synthesis of T helper 2-type cytokines at the maternal-fetal interface. J Immunol. 1993;151:4562–73.
18. Dosiou C, Giudice L. Neural killer cells in pregnancy and recurrent pregnancy loss: endocrine and immunologic perspectives. Endocr Rev. 2005;261:44–62.
19. Alimohamadi S, Javadian P, Gharedaghi MH, Javadian N, Alinia H, Khazardoust S, Borna S, Hantoushzadeh S. Progesterone and threatened abortion: a randomized clinical trial on endocervical cytokine concentrations. J Reprod Immunol. 2013;98:52–60.
20. Aisemberg J, Vercelli CA, Bariani MV, Billi SC, Wolfson ML, Franchi AM. Progesterone is essential for protecting against LPS-induced pregnancy loss. LIF as a potential mediator of the anti-inflammatory effect of progesterone. PLoS One. 2013;8(2):e56161. doi:10.1371/journal.pone.0056161.
21. Raghupathy R, Al Mutawa E, Makhseed M, Azizieh F, Szekeres-Bartho J. Modulation of cytokine production by dydrogesterone in lymphocytes from women with recurrent miscarriage. BJOG. 2005;112(8):1096–101.
22. Hudić I, Szekeres-Bartho J, Fatušić Z, Stray-Pedersen B, Dizdarević-Hudić L, Latifagić A, Hotić N, Kamerić L, Mandžić A. Dydrogesterone supplementation in women with threatened preterm delivery–the impact on cytokine profile, hormone profile, and progesterone-induced blocking factor. J Reprod Immunol. 2011;92(1–2):103–7.
23. Retamales I, Carrasco I, Troncoso JL, Las Heras J, Devoto L, Vega M. Morpho-functional study of human luteal cell subpopulations. Hum Reprod. 1994;9:591–6.
24. Fraser HM, Lunn SF. Regulation and manipulation of angiogenesis in the primate corpus luteum. Reproduction. 2001;121:3554–62.
25. Suzuki T, Sasano H, Takaya R, et al. Cyclic changes of vasculature and vascular phenotypes in normal human ovaries. Hum Reprod. 1988;13:953–9.
26. Schams D, Berisha B. Regulation of corpus luteum function in cattle-an overview. Reprod Domest Anim. 2004;39:241–51.

27. Fatemi HM, Bourgain C, Donoso P, et al. Effect of oral administration of dydrogesterone versus vaginal administration of natural micronized progesterone on thee secretory transformation of endometrium and luteal endocrine profile in patients with premature ovarian failure: a proof of concept. Hum Reprod. 2007;22:1260–3.
28. Poenzias AS. Luteal phase support. Fertil Steril. 2002;77:318–23.
29. McNatty KP, Smith DM, Makris A, et al. The microenvironment of the human antral follicle: interrelationships among the steroid levels in antral fluid, the population of granulosa cells, and the status of the oocyte in vivo and in vitro. J Clin Endocrinol Metab. 1979;49:851–60.
30. McNatty KP, Makris A, DeGrazia C, et al. The production of progesterone, androgens, and estrogens by granulosa cells, thecal tissue, and stromal tissue from human ovaries in vitro. J Clin Endocrinol Metab. 1979;49:687–99.
31. Norris RP, Freudzon M, Mehlmann LM, et al. Luteinizing hormone causes MAP minase-dependent phosphorylation and closure of connexin 43 gap junctions in mouse ovarian follicles: one of two paths to meiotic resumption. Development. 2008;135:3229–38.
32. Miller WL. Mechanizm of StAR's regulation of mitochondrial cholesterol import. Mol Cell Endocrinol. 2007;265–6:46–50.
33. Kiriakidou M, McAllister JM, Sugawara T, Strauss 3rd JF. Expression of steroidogenic acute regulatory protein (StAR) in the human ovary. J Clin Endocrinol Metab. 1986;81:4122–8.
34. Noyes RW, Hertig AT, Rock J. Dating the endometrial biopsy. Fertil Steril. 1950;1:3–10.
35. McNeely MJ, Soules MR. The diagnosis of luteal phase deficiency: a critical review. Fertil Steril. 1989;51:582–7.
36. Noyes RW, Haman JO. Accuracy of endometrial dating: correlation of endometrial dating with basal body temperature and menses. Fertil Steril. 1953;4:504–9.
37. Peters AJ, Lloyd RP, Coulam CB. Prevalence of out-of-phase endometrial biopsy specimens. Am J Obstet Gynecol. 1992;166:1738–41.
38. Abraham GE, Maroulis GB, Marshall JR. Evaluation of ovulation and corpus luteum function using measurements of plasma progesterone. Obstet Gynecol. 1974;44:522–7.
39. Huybayter ZR, Muasher SJ. Luteal supplementation in vitro fertilization: more questions than answers. Fertil Steril. 2008;89:749–58.
40. Huytchinson-Williams KA, Lunefeld B, Diamond MP, et al. Human chorionic gonadotropin, estradiol, and progesterone profiles in conception and non-conception cycles in an in vitro fertilization program. Fertil Steril. 1989;52:441–5.
41. Check JH. Progesterone therapy versus follicle maturing drugs possible opposite effects on embryo implantation. Clin Exp Obstet Gynecol. 2002;29:5–10.
42. Macklon NS, Fauser BC. Impact of ovarian hyper-stimulation on the luteal phase. J Reprod Fertil Suppl. 2000;55:101–8.
43. Erdem A, Erdem M, Atmaca S, Guler I. Impact of luteal phase support on pregnancy rates in intrauterine insemination cycles; a prospective randomized study. Fertil Steril. 2009;91: 2508–13.
44. Forman RG, Eychenne B, Nessmann C, et al. Assessing the early luteal phase in-vitro fertilization cycles: relationships between plasma steroids, endometrial receptors, and endometrial histology. Fertil Steril. 1989;51:310–6.
45. Li TC, Tuckerman EM, Laird SM. Endometrial factors in recurrent miscarriage. Hum Reprod Update. 2002;1:43–52.
46. Schweikert A, Rau T, Berkholz A, et al. Association of progesterone receptor polymorphism with recurrent abortions. Eur J Obstet Gynecol Reprod Biol. 2004;113:67–72.
47. Resenberg SM, Luciano AA, Riddick DH. The luteal phase defect: the relative frequency of, and encouraging response to, treatment with vaginal progesterone. Fertil Steril. 1980; 34:17–20.
48. Hutchinson-Williams KA, DeCherney AH, Lavy G, et al. Luteal rescue in vitro fertilization-embryo transfer. Fertil Steril. 1990;53:495–500.
49. Pritts EA, Atwood AK. Luteal phase support in infertility treatment: a meta-analysis of the randomized trials. Hum Reprod. 2002;17:2287–99.

50. Nosarka S, Kruger T, Siebert I, Grove D. Luteal phase support in in vitro fertilization: meta-analysis of randomized trials. Gynecol Obstet Invest. 2005;60:67–74.
51. van der Linden M, Buckingham K, Farquhar C, et al. Luteal phase support for assisted reproduction cycles. Cochrane Data base Syst rev 2011;5:CD009154.
52. Ludwig M, Diedrich K. Evaluation of an optimal luteal phase support protocol in IVF. Acta Obstet Gynecol Scand. 2001;80:452–66.
53. Kol S, Humaidan P, Itskovitz-Eldor J. GnRH agonist ovulation trigger and hCG-based, progesterone-free luteal support: a proof of concept study. Hum Reprod. 2011;26:2874–7.
54. Devroey P, Palermo G, Bourgain C, et al. Progesterone administration in patients with absent ovaries. Int J Fertil. 1989;34:188–93.
55. Ludwig M, Schwartz P, Babahan B, et al. Luteal phase support using either Crinone 8% or Utrogest: results of a prospective randomized study. Eur J Obstet Gynecol Reprod Biol. 2002;103:48–52.
56. Polyzos NP, Cl M, Papanikolau EG, et al. Vaginal progesterone gel for luteal phase support in IVF/ICSI cycles: a meta-analysis. Fertil Steril. 2010;94:2083–7.
57. Silverberg KM, Vaughn TC, Hansard LJ, et al. Vaginal (Crinone 8%) gel vs. intramuscular pro progesterone in oil for luteal phase support in in vitro fertilization: a large prospective trial. Fertil Steril. 2012;97:344–8.
58. Gorkemli H, Ak D, Akyurek C, et al. Comparison of pregnancy outcomes, of progesterone or progesterone + estradiol for luteal phase support in IFSI-ET cycles. Gynecol Obstet Invest. 2004;58:140–4.
59. Kolibianakis EM, Venetis CA, Papanikolau EG, et al. Estrogen addition to progesterone for luteal phase support in cycles stimulated with GnRH analogues and gonadotrophins for IVF: a systematic review and meta-analysis. Hum Reprod. 2008;23:1346–54.
60. Pirard C, Donnez J, Loumaye E. GnRH agonist as luteal phase support in assisted reproduction technique cycles: results of a pilot study. Hum Reprod. 2006;21:1894–900.
61. Kykrou D, Kolibianakis EM, Fatemi HM, et al. Increased live birth rates with GnRH agonist addition for luteal support in ICSI/IVF cycles: a systematic review and meta-analysis. Hum Reprod Update. 2011;17:734–40.
62. Queisser-Luft A. Dydrogesterone use during pregnancy: overview of birth defects reported since 1977. Early Hum Dev. 2009;85:375–7.
63. King RJ, Whitehead MI. Assessment of the potency of orally administered progestins in women. Fertil Steril. 1986;46:1062–6.
64. Chakravarty BN, Shirazee HH, Dam P, et al. Oral dydrogesterone versus intravaginal micronised progesterone as luteal phase support in assisted reproductive technology (ART) cycles: results of a randomized study. J Steroid Biochem Mol Biol. 2005;97:416–20.
65. Iwase A, Ando H, Toda S, et al. Oral progestogen versus intramuscular progesterone for luteal support after assisted reproductive technology treatment: a prospective randomized study. Arch Gynecol Obstet. 2008;277:319–24.
66. Patki A, Pawar VC. Modulating fertility outcome in assisted reproductive technologies by the use of dydrogesterone. Gynecol Endocrinol. 2007;23 Suppl 1:68–72.

Chapter 4
Progestogens in Threatened Miscarriage

Howard J.A. Carp

1 Introduction

Threatened miscarriage is defined by the National Library of Medicine, Medical Subject Headings (2012 MeSH), as bleeding during the first 20 weeks of pregnancy while the cervix is closed. It is the most common complication in pregnancy, occurring in 20 % of all pregnancies. The condition may progress to miscarriage in approximately one half of cases [1, 2], or may resolve. There are problems of definition, as the bleeding may include anything from spots of blood to potentially fatal shock. The treating physician is faced with the question whether any treatment can effectively prevent the pregnancy from being miscarried. Progestational agents have been prescribed since the 1950s in order to prevent miscarriages. There is much theoretical data to support the use of progestogens. Progestogens enhance implantation, affect the cytokine balance, inhibit natural killer cell activity at the feto-maternal interface, inhibit the release of arachidonic acid, prevent myometrial contractility and prevent cervical dilatation. Activation of progesterone receptors on lymphocytes results in the synthesis of a protein known as progesterone-induced blocking factor (PIBF) [3]. PIBF favours the production of asymmetric, pregnancy-protecting antibodies.

Indeed it has often been reported that a progesterone lack may lead to miscarriage [4, 5]. Lutectomy prior to 7 weeks causes miscarriage [6]. Mifepristone blocks the progesterone receptor, leading to fetal death and placental separation. Therefore, the question arises whether progesterone supplementation should be used, and if so, which progestogen. In addition, the results of treatment need to be compared to the natural history of the condition. There are a number of trials in the literature

H.J.A. Carp, M.B.B.S., F.R.C.O.G. (✉)
Department of Obstetrics & Gynecology, Sheba Medical Center, Tel Hashomer 52621, Israel

Sackler School of Medicine, Tel Aviv University, Tel Aviv, Israel
e-mail: carp@netvision.net.il

H.J.A. Carp (ed.), *Progestogens in Obstetrics and Gynecology*,
DOI 10.1007/978-3-319-14385-9_4

comparing progestogens to no treatment or placebo, stretching back over 50 years. This chapter will summarise the results of treatment, pitfalls and confounding factors.

2 Natural History

There are problems in determining the natural history of threatened miscarriage as the definition takes no account of the severity of bleeding. However, if the condition is assumed to be homogeneous, it has been reported that threatened miscarriage may progress to abortion in approximately 50 % of cases [1, 2], or may resolve. However, in the older literature, there was no ultrasound performed in order to detect the fetal heartbeat. In many cases, bleeding may have occurred after fetal death. Table 4.1 shows the subsequent history after detection of a fetal heartbeat in a number of observational studies. As can be seen 8.7 % of threatened miscarriages subsequently terminated in miscarriage. 92.3 % continued developing.

Weiss et al. [7] enrolled patients into a database on presenting with a viable embryo at 10–14 weeks. If the patient reached 10–14 weeks, the chance of miscarrying prior to 24 weeks was 1–2 %. Another source of ascertainment of the miscarriage rate after threatened miscarriage is from the control group of randomised trials of progestogens in threatened miscarriage. These can be seen in Figs. 4.2 and 4.3. In these nine studies the miscarriage rate was higher. 129 pregnancies of 519 resulted in miscarriage (24.9 %). However, these figures are not corrected for the presence of a fetal heartbeat prior to enrollment in the various trials.

In recurrent miscarriage, vaginal bleeding is a common complication occurring in 50 of 162 women in Reginald's series [8] and 50 of 102 patients in the author's series [9]. The reason for this bleeding remains unclear. 75 % of habitual abortions are blighted ova [9]. However, when the pregnancy succeeds and there is a live embryo within the uterus, bleeding still occurs in 40–50 % of patients. The likelihood of a pregnancy loss after the detection of fetal heartbeats as 69/359 (14.2 %) in Li et al.'s series [10] and 22.7 % of 185 study patients with multiple spontaneous abortions in Laufer et al.'s [11] series.

Table 4.1 Prognosis after detection of fetal heartbeat (observational studies)

Series	Miscarriages	Total	Proportion miscarrying (%)
Tongsong et al. [56]	14	255	5.6
Tannirandorn et al. [57]	3	87	3.4
Falco et al. [58]	23	149	15.4
Bennett et al. [38]	48	516	9.3
Subchorionic hematoma			
Total	88	1,007	8.7

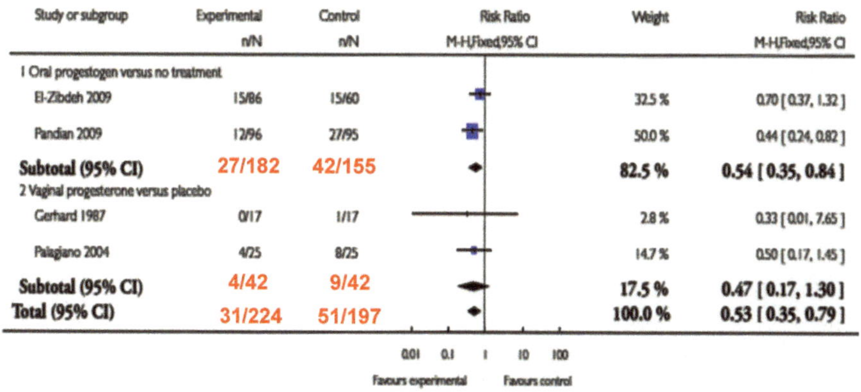

Modified from Wahabi et al (22). Original reproduced by permission of John Wiley and Sons

Fig. 4.1 Metaanalysis of RCT's on progestogen supplementation. Modified from Wahabi et al. [22]. Original reproduced by permission of John Wiley and Sons

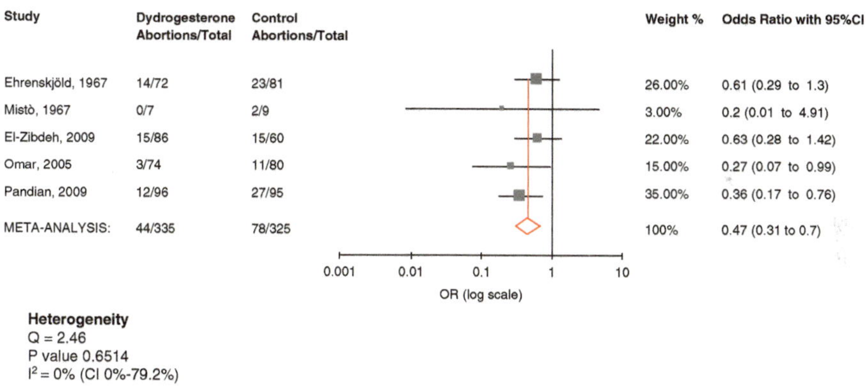

Reporoduced by permission from Informa Healthcare.

Fig. 4.2 Metaanalysis of dydrogesterone in threatened miscarriage [32]. Reproduced by permission from Informa Healthcare

3 Progesterone Levels

Serum progesterone levels have been used to make prognoses about the continued development of pregnancy and even to diagnose pregnancy loss. The lowest progesterone level to be associated with a viable pregnancy was 5.1 ng/ml in Stovall et al.'s [12] series. A single progesterone level ≥25 ng/ml was associated with a 97 % likelihood of viable pregnancy [12]. Al-Sebai et al. [5] summarised 358 threatened

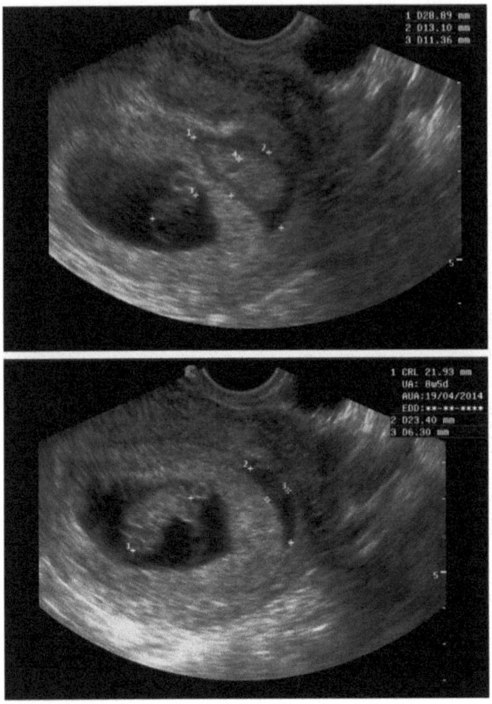

Fig. 4.3 Subchorionic hematoma at 7 and 9 weeks

abortions <18 weeks, a single progesterone level ≤ 45 nmol/l (14 ng/ml) was reported to differentiate between aborting and ongoing pregnancies (sensitivity 87.6 %, specificity 87.5 %). Serum progesterone levels of less than ≤12 ng/ml. were associated with an increased risk of miscarriage in Arck et al.'s [13] series.

The mechanisms by which luteal phase deficiency occur are unclear but could be associated with decreased FSH levels in the follicular phase (it has been suggested that luteal phase deficiency originates as a preovulatory event). Alternatively there may be an abnormal pattern of LH secretion. Luteal phase deficiency may be secondary to abnormal follicle formation, associated with poor oocyte quality, or a decreased response to progesterone by the endometrium [14]. However, there are pitfalls to using serum progesterone levels as a predictive marker of miscarriage, or for determining the need for progesterone supplementation. Progesterone secretion is pulsatile. Blood may be drawn at a pulse peak or nadir. These may vary tenfold [15]. Hormone levels may be normal, but histology abnormal due to deficiency of progesterone receptors. As with other presumptive causes of abortion, low hormone levels may be a result of abortion rather than its cause. In the blighted ovum or after embryonic death, there is no villous circulation. Trophoblastic failure after villous circulatory failure results in low hCG levels. If hCG does not stimulate the corpus luteum, progesterone levels will fall explaining the mechanism of expulsion, but not necessarily that of embryonic death, or the cause of abortion. Therefore in threatened miscarriage, diagnosis and treatment should be empirical rather than based on progesterone levels.

4 Confounding Factors

The results of progestogen treatment may be confounded as threatened miscarriage, may be due to separation of the placenta in a normal embryo, or a defence mechanism to prevent the continued development of an abnormal embryo, including abnormalities which are incompatible with life. The most important confounding factors are embryonic structural malformations, or chromosomal aberrations. In missed abortions 200 of 233 embryos have been reported to be structurally abnormal on embryoscopy [16]. These defects included: anencephaly, encephalocele, spina bifida, syndactyly, pseudo-syndactyly, polydactyly, cleft hand and cleft lip. Without embryoscopy these embryos would not have been diagnosed, and the patient might have been treated empirically with progesterone. If included in a trial, the results would be skewed in favour of a negative effect. However, embryoscopy is advanced technique, which is not usually available. Additionally, 70 % of miscarriages are blighted ova. Therefore, it is impossible to tell if a rudimentary embryo may have been structurally abnormal. At present, ultrasound is not sufficiently sensitive to diagnose these very early anomalies.

Up to 60 % of sporadic miscarriages [17–19] may be caused by chromosomal aberrations in the embryo. The most common aberrations which are incompatible with life include 16 trisomy and polyploidy. Additionally, with the introduction of comparative genomic hybridization (CGH) a further 15 % of so called normal embryos on conventional karyotyping, have been diagnosed with genetic aberrations [20, 21].

Progesterone cannot correct chromosomal aberrations. Unfortunately, the abortus is not usually karyotyped, even less are subjected to CGH microarray analysis. Hence, it is unclear whether miscarriage after supplemental progesterone is due to failure of treatment or confounding of the results by fetal chromosomal aberrations.

Before a trial of progesterone can be said to be conclusive, other predictive factors should be taken into account.

5 Effect of Progestogen Supplementation

Progestogens have been used since the 1950s to prevent threatened miscarriage terminating as miscarriage. It has been difficult to prove whether progestogens are effective, due to the generally good natural history, and the effect of confounding factors. Recently, two systematic reviews have been performed. Wahibi et al. [22] carried out an analysis of two trials of oral dydrogesterone compared to placebo [23, 24] and two trials of vaginal progesterone [25, 26] (See Fig. 4.1). The overall figures showed a statistically significant benefit, OR = 0.53 (CI 0.35–0.79) in favour of progestogen supplementation. However, the analysis consisted of only four studies. The authors concluded, "The analysis was limited by the small number and the poor methodological quality of eligible studies, and the small number of the participants, which limit the power of the meta-analysis and hence of the conclusion". It is interesting to note that in the women who were treated with vaginal progesterone the treatment was not statistically effective in reducing miscarriage when compared to placebo (RR=0.47 95 % CI, 0.17–1.30) whereas oral progestogen (dydrogesterone) was effective (RR=0.54 CI 0.35–0.84). In the light of Wahibi et al.'s [22] metaanalysis, two new guidelines were released regarding threatened miscarriage. In the U.K. the National Institute for Health and Care Excellence (NICE) [27] produced a guideline in 2012 which stated, that "data from a meta-analysis of several small studies suggest that progestogens are better than placebo". The Royal Australian and New Zealand College of Obstetricians and/gynaecologists guideline [28] stated that, "Miscarriage was significantly less likely to occur on progestins than placebo or no treatment (risk ratio 0.53; 95 % CI 0.35–0.79)". Both guidelines stated, that further research was necessary to confirm the results on a larger number of patients.

The author tried to enlarge the number of patients available for analysis by relaxing the strict methodological rules required by the Cochrane database for inclusion in meta-analyses. A literature search was performed for in September 2010 for all papers available at that time in EMBASE and Ovid MEDLINE® that fulfilled the following criteria: original contributions with product name 'Progestogen, Progestin, Progesterone, Micronised progesterone Duphaston' or 'dydrogesterone', reports limited to clinical human data excluding reviews, case reports and editorials in any language. All articles considered were investigator initiated trials and published in the scientific literature.

No additional relevant trials were found for micronized or intramuscular progesterone, to those assessed in Wahibi et al.'s [22] metaanalysis. However, three additional trials of oral dydrogesterone were considered eligible for inclusion in a systematic review [29–31]. Carp's [32] metaanalysis included five randomized studies. 660 patients were eligible for analyses of pregnancy maintenance. The results (Fig. 4.2) showed that the effect of dydrogesterone on the risk of miscarriage in women with threatened miscarriage appeared to be substantial. There was a statistically significant reduction in the odds ratio for miscarriage after dydrogesterone compared to standard care of 0.47 (CI 0.31–0.7). The 24 % miscarriage rate in control women (78/325) was reduced to 13 % (44/335) after dydrogesterone administration (11 % absolute reduction in the miscarriage rate). There has since been a further trial of vaginal micronised progesterone. Yassaee et al [33] have reported on the results of 60 patients treated with threatened miscarriage who were treated with vaginal micronised progesterone. Although there was a trend towards an improved prognosis, the results were not significantly better than in the control group.

5.1 Evidence of Effect

There are two metaanalyses in the literature which show progestogens to lower the risk of threatened miscarriage developing to miscarriage. In both metaanalyses the effect of dydrogesterone was statistically significant (Grade I). There a dearth of evidence about 17 hydroxyprogesterone acetate or caproate by intramuscular injection, as only one small trial of been reported to restore levels of soluble TNF receptors to normal in women with recurrent pregnancy loss [34]. However, Shearman and Garrett [35] found 17 hydroxyprogesterone caproate to have no beneficial effect in threatened miscarriage. Additionally, when myometrial tissues were suspended in organ chambers and exposed to varying concentrations of progesterone or 17 hydroxyprogesterone caproate [36] progesterone significantly inhibited spontaneous contractility, but 17 hydroxyprogesterone caproate actually stimulated contractility. It is therefore not recommended for threatened miscarriage. Unfortunately, only two trials on vaginal progesterone were analysed in Wahibi et al.'s [22] meta-analysis. There were only 84 patients in both trials. Although there was a reduction in the number of miscarriages progressing to miscarriage, both trials were not sufficiently powered to reach statistical significance. Palagiano et al. [26], and Yassaee et al, [33] used micronised progesterone. Neither trial reached statistical significance.

Needless to state that no studies control for karyotypic aberrations or predictive factors.

5.2 *Subgroups of Threatened Miscarriage*

5.2.1 Subchorionic Hematoma

Subchorionic hematoma is a common occurrence in threatened miscarriage, being found in approximately 18 % of all cases of bleeding during the first trimester [37]. There is one observational study on the natural history of subchorionic hematoma in threatened abortion after detection of the fetal heart [38]. The incidence of miscarriage was 8.9 %, similar to other cases of threatened miscarriage. The authors concluded that, "For women with a subchorionic hematoma that is sonographically identified, fetal outcome is dependent on size of the hematoma, maternal age, and gestational age". A metaanalysis by Tuuli et al. [39] which assessed trials in which the presence of a fetal heart was not identified, included 1,735 women with a subchorionic hematoma. 17.6 % of pregnancies progressed to miscarriage. Figure 4.3 shows a patient with a subchorionic hematoma from the author's series.

There are two trials of progestogens in subchorionic hematoma. Both are open labelled observational studies. In the first study, Pelinescu-Onciul et al. [40] treated 125 women with micronized progesterone 600 mg/day. 18.7 % of pregnancies terminated in miscarriage. In the second study [41], 100 women, with threatened miscarriage and a viable embryo received dydrogesterone. There were 93 live births and 7 miscarriages. The difference in results was significantly better in the dydrogesterone group (RR = 2.04 CI = 1.05–3.97). However, these results should be treated with caution due to the methodological flaws of comparing two separate cohorts of patients who were not randomized.

5.2.2 Threatened Miscarriage After Recurrent Miscarriage

As stated above, vaginal bleeding occurs in 40–50 % of recurrently miscarrying women, even when the pregnancy is viable. Progestogen supplementation has been used in recurrent miscarriage with varying degrees of success, as described in the next chapter. As in threatened miscarriage, treatment can only affect a live embryo, or an embryo at a stage prior to 5.5 weeks (usually the earliest that a fetal heart can be detected.). A thorough search of the literature failed to find any reports of progestogen use to prevent threatened miscarriage developing to miscarriage in women with previous recurrent miscarriage. It is possible that progestogen supplementation prevents threatened miscarriage from occurring.

6 Safety and Side Effects

Safety and side effects are always a major worry when drugs are used in pregnancy. The major concerns about safety have centred on the effect of progestogens on the androgen receptor. The original progestogens were derivatives of 19 nor testosterone. These compounds led to varying degrees of masculinization of female fetuses. The masculinization included clitoral enlargement, and varying degrees of labial fusion. However, these side effects are not seen with other progestins. On the contrary, progesterone itself has anti-androgenic effects and has been reported to lead to hypospadias. In a case–control study, a relationship was reported between the use of progesterone and isolated hypospadias in two uncontrolled observational studies [42, 43]. Carmichael et al. [44], studied the risk of hypospadias and periconceptional progestin intake. Progestin-related hypospadia was reported by 42 (8.4 %) case mothers and 31 (2.4 %) control mothers, for intakes from 4 weeks before conception to 14 weeks of pregnancy. The crude odds ratio for progestin use at any time was 3.7 (95 % CI 2.3–6.0). Additionally, Rock et al. [45] reported one case of undescended testis and one case of meningomyelocele in 93 women treated with progesterone in the first trimester. Check et al. [46] found two cardiovascular malformations, omphalocele, hydrocephalus and club foot (with cleft palate) in 382 women exposed to either progesterone or 17-α hydroxy-progesterone. These studies had no control group.

Progesterone itself, has been reported as safe, with no increase in the frequency of congenital anomalies [47, 48]. In 1999, the US FDA classified micronized progesterone, as a category B drug. It appears that progesterone treatment does not increase the risk of nongenital birth defects.

However, there are side effects with micronized progesterone. If given orally, the hormone is degraded in the liver. In miscarriage, there is no data on efficacy, and there is extreme variability in plasma concentrations [49]. Side-effects include nausea, headache and sleepiness. If micronized progesterone is administered vaginally, hepatic metabolism is avoided. The suppositories are not painful, and side effects are few. However, there are problems with patient compliance, as vaginal tablets are not acceptable in some cultures. In addition, vaginal administration is uncomfortable if there is bleeding or discharge and the suppositories may be washed out if bleeding is severe. There are also side effects concerning patient comfort. The internet contains numerous reports and images of side effects such as excessive and irritant vaginal discharge, vulval edema, irritation etc. These side effects are not usually mentioned in the professional literature.

17 hydroxyprogesterone by intramuscular injection has numerous side effects including: Extreme pain, swelling, itching and other local reactions at the injection site, abscesses formation, hypersensitivity reactions, cough, dyspnea, tiredness, dizziness, genital itching, and increased risk of gestational diabetes, mood swings, headaches, bloating, abdominal pain, perineal pain, constipation, diarrhea, nausea, vomiting, joint pain, depression, decreased sex drive, nervousness, sleepiness, breast enlargement, breast pain, dysuria, polyuria, UTI, vaginal discharge, fever, flu-like symptoms, back pain, leg pain, sleep disorder, upper respiratory infection, asthma, acne and pruritus. There have been concerns regarding the vehicle, castor oil, which may induce labor by stimulating release of prostaglandins [50, 51]. Three clinical studies in singleton pregnancies have all shown increased risk of miscarriage compared to placebo [52–54]. 17 hydroxy progesterone acetate or caproate are therefore not recommended for threatened miscarriage.

In order to assess the safety of dydrogesterone, all 22 studies originally considered for Carp's [32] metaanalysis on dydrogesterone were reviewed. The follow up data on 1,380 patients suggests that the side effects including birth defects are minimal. Additionally, a review of birth defects associated with dydrogesterone use during pregnancy [55] concluded that clinical experience with dydrogesterone provided no evidence of a causal link between maternal use during pregnancy and birth defects. It is estimated that between 1977 and 2005 approx. 38 million women treated with dydrogesterone, and more than 10 million fetuses exposed. There also seem to be no major side effects in the mother.

7 Conclusions

In the light of observational studies showing that 90 % of threatened miscarriages continue developing after detection of a fetal heartbeat, it is difficult to give recommendations. If the embryo is chromosomally abnormal, it will be miscarried whatever treatment is advised. Additionally, there is no accurate diagnosis of progesterone deficiency. However, progestogen administration has been shown to reduce the number of threatened miscarriages developing to miscarriage. There are three trials of micronized progesterone to date, and these studies were underpowered to show statistical significance. 17 hydroxy progesterone caproate seems to increase the likelihood of miscarriage, and is therefore not recommended. Dydrogesterone however, has been assessed on 660 patients in a systematic review, and the effect seems to be substantial, reducing the odds ratio for miscarriage by 47 %.

References

1. Farrell T, Owen P. The significance of extra chorionic membrane separation in threatened miscarriage. BJOG. 1996;103:926–8.
2. Everett C. Incidence and outcome of bleeding before the 20th week of pregnancy: prospective study from general practice. BMJ. 1997;315:32–4.
3. Szekeres-Bartho J, Barakonyi A, Par G, Poigar B, Palkovics T, Szereday L. Progesterone as an immunomodulatory molecule. Int Immunopharmacol. 2001;1:1037–48.
4. Stovall TG, Ling FW, Andersen RN, Buster JE. Improved sensitivity and specificity of a single measurement of serum progesterone over serial quantitative beta-human chorionic gonadotrophin in screening for ectopic pregnancy. Hum Reprod. 1992;7:723–5.
5. al-Sebai MA, Kingsland CR, Diver M, Hipkin L, McFadyen IR. The role of a single progesterone measurement in the diagnosis of early pregnancy failure and the prognosis of fetal viability. Br J Obstet Gynaecol. 1995;102:364–9.
6. Csapo AI, Pulkkinen MO. Indispensability of the human corpus luteum in the maintenance of early pregnancy: lutectomy evidence. Obstet Gynecol Surv. 1978;3:69–81.
7. Weiss JL, Malone FD, Vidaver J, Ball RH, Nyberg DA, Comstock CH, et al. Threatened abortion: a risk factor for poor pregnancy outcome, a population-based screening study. Am J Obstet Gynecol. 2004;190:745–50.
8. Reginald PW, Beard RW, Chapple J, Forbes PB, Liddell HS, Mowbray JF, Underwood JL. Outcome of pregnancies progressing beyond 28 weeks gestation in women with a history of recurrent miscarriage. Brit J Obstet Gynaecol. 1987;94:643–8.
9. Carp HJA, Toder V, Mashiach S, et al. Recurrent miscarriage: a review of current concepts, immune mechanisms, and results of treatment. Obst Gynecol Surv. 1990;45:657–69.
10. Li TC, Makris M, Tomsu M, Tuckerman E, Laird S. Recurrent miscarriage: aetiology, management and prognosis. Hum Reprod Update. 2002;8:463–81.
11. Laufer MR, Ecker JL, Hill JA. Pregnancy outcome following ultrasound-detected fetal cardiac activity in women with a history of multiple spontaneous abortions. J Soc Gynecol Investig. 1994;1(2):138–42.
12. Stovall TG, Ling FW, Carson SA, Buster JE. Serum progesterone and uterine curettage in differential diagnosis of ectopic pregnancy. Fertil Steril. 1992;57:456–7.
13. Arck PC, Rücke M, Rose M, Szekeres-Bartho J, Douglas AJ, Pritsch M, et al. Early risk factors for miscarriage: a prospective cohort study in pregnant women. Reprod Biomed Online. 2008;17(1):101–13.
14. Tuckerman E, Laird SM, Stewart R, Wells M, Li TC. Markers of endometrial function in women with unexplained recurrent pregnancy loss. Hum Reprod. 2004;19:196–205.
15. Abraham GE, Maroulis GB, Marshall JR. Evaluation of ovulation and corpus luteum function using measurements of plasma progesterone. Obstet Gynecol. 1974;44:522–5.
16. Philipp T, Philipp K, Reiner A, Beer F, Kalousek DK. Embryoscopic and cytogenetic analysis of 233 missed abortions: factors involved in the pathogenesis of developmental defects of early failed pregnancies. Hum Reprod. 2003;18:1724–32.
17. Boue J, Boue A, Lazar P. The epidemiology of human spontaneous abortions with chromosomal anomalies. In: Blandau RJ, editor. Aging gametes. Karger & Co: Basel; 1975. p. 330.
18. Sanchez JM, Franzi L, Collia F, De Diaz SL, Panal M, Dubner M. Cytogenetic study of spontaneous abortions by transabdominal villus sampling and direct analysis of villi. Prenat Diag. 1999;19:601–3.
19. Stein Z. Early fetal loss. Birth Def. 1981;17:95–9.
20. Miller DT, Adam MP, Aradhya S, Biesecker LG, Brothman AR, Carter NP, et al. Consensus statement: chromosomal microarray is a first-tier clinical diagnostic test for individuals with developmental disabilities or congenital anomalies. Am J Hum Genet. 2010;86:749–64.
21. Rajcan-Separovic E, Diego-Alvarez D, Robinson WP, Tyson C, Qiao Y, Harvard C, Fawcett C, Kalousek D, Philipp T, Somerville MJ, Stephenson MD. Identification of copy number

variants in miscarriages from couples with idiopathic recurrent pregnancy loss. Hum Reprod. 2010;25:2913–22.

22. Wahabi HA, Fayed AA, Esmaeil SA, Al Zeidan RA. Progestogen for treating threatened miscarriage. Cochrane Database Syst Rev 2011;(12):CD005943.
23. El-Zibdeh MY, Yousef LT. Dydrogesterone support in threatened miscarriage. Maturitas. 2009;65 Suppl 1:S43–6.
24. Pandian RU. Dydrogesterone in threatened miscarriage: a Malaysian experience. Maturitas. 2009;65 Suppl 1:S47–50.
25. Gerhard I, Gwinner B, Eggert-Kruse W, Runnebaum B. Double-blind controlled trial of progesterone substitution in threatened abortion. Biol Res Pregnancy Perinatol. 1987;8:26–34.
26. Palagiano A, Bulletti C, Pace MC, De Ziegler D, Cicinelli E, Izzo A. Effects of vaginal progesterone on pain and uterine contractility in patients with threatened abortion before twelve weeks of pregnancy. Ann N Y Acad Sci. 2004;1034:200–10.
27. National Institute for Health and Care Excellence. Nice clinical Guideline No. 154, Ectopic pregnancy and miscarriage. 2012; http://www.nice.org.uk/guidance/cg154
28. Royal Australian and New Zealand College of Obstetricians and Gynaecologists. Progesterone support of the luteal phase and in the first trimester. 2013; www.ranzcog.edu.au
29. Ehrenskjöld ML, Bondo B, Weile F. Treatment of threatened abortion with dydrogesterone. Ugeskr Laeger. 1967;129:1678–9 [Article in Danish].
30. Mistò A. Experiences with 6-dehydro-retroprogesterone int the treatment of placental insufficiency. Ann Ostet Ginecol Med Perinat. 1967;89:102–12 [Article in Italian].
31. Omar MH, Mashita MK, Lim PS, Jamil MA. Dydrogesterone in threatened abortion: pregnancy outcome. J Steroid Biochem Mol Biol. 2005;97:421–5.
32. Carp H. A systematic review of dydrogesterone for the treatment of threatened miscarriage. Gynecol Endocrinol. 2012;28:983–90.
33. Yassaee F, Shekarriz-Foumani R, Afsari S, Fallahian M.The effect of progesterone suppositories on threatened abortion: a randomized clinical trial. J Reprod Infertil. 2014; 15:147–51.
34. Chernyshov VP, Vodyanik MA, Pisareva SP. Lack of soluble TNF-receptors in women with recurrent spontaneous abortion and possibility for its correction. Am J Reprod Immunol. 2005;54:284–91.
35. Shearman RP, Garrett WJ. Double-blind study of effect of 17- hydroxyprogesterone caproate on abortion rate. Br Med J. 1963;1(5326):292–5.
36. Ruddok NK, Shi SQ, Jain S, Moore G, Hankins GDV, Romero R, Garfield RE. Progesterone, but not 17-alpha-hydroxyprogesterone caproate, inhibits human myometrial contractions. Am J Obstet Gynecol. 2008;199(391):e1–7.
37. Sauebrei EE. Early pregnancy: pre-embryonic and embryonic periods. In: Sauebrei EE, Nguyen KT, Nolan RL, editors. A practical guide to ultrasound in obstetrics and gynecology. Philadelphia, PA: Lippincott-Raven Publishers; 1998. p. 122–31.
38. Bennett GL, Bromley B, Lieberman E, Benacerraf BR. Subchorionic hemorrhage in first-trimester pregnancies: prediction of pregnancy outcome with sonography. Radiology. 1996;200(3):803–6.
39. Tuuli MG, Norman SM, Odibo AO, Macones GA, Cahill AG. Perinatal outcomes in women with subchorionic hematoma: a systematic review and meta-analysis. Obstet Gynecol. 2011;117:1205–12.
40. Pelinescu-Onciul D, Radulescu-Botica R, Steriu M, Cheles C, Varlas V. Terapia cu progesteron micronizat a hematoamelor eciduale. Infomedica. 1999;2S:32–5.
41. Pelinescu-Onciul D. Subchorionic hemorrhage treatment with dydrogesterone. Gynecol Endocrinol. 2007;23 Suppl 1:77–81.
42. Kallen B, Castilla EE, Robert E, Lancaster PAL, Kringeelbach M, Mutchinick O, et al. An international case–control study on hypospadias. The problem with variability and the beauty of diversity. Eur J Epidemiol. 1992;8:256–63.
43. Kallen B, Martinez-Frias ML, Castilla EE, Robert E, Lancaster PAL, Kringelbach M, et al. Hormone therapy during pregnancy and isolated hypospadias: an international case–control study. Int J Risk Saf Med. 1992;3:183–98.

44. Carmichael SL, Shaw GM, Laurent C, Croughan MS, Olney RS, Lammer EJ. Maternal progestin intake and risk of hypospadias. Arch Pediatr Adolesc Med. 2005;159:957–62.
45. Rock JA. Fetal malformations following progesterone therapy during pregnancy: a preliminary report. Fertil Steril. 1985;44:17–9.
46. Check JH. The risk of fetal anomalies as a result of progesterone therapy during pregnancy. Fertil Steril. 1986;45:575–7.
47. Bartholomeusz RK, Bruce NW. Effects of maternal progesterone supplementation on fetal, placental and corpus luteal weights in the rat. Biol Reprod. 1976;15:84–9.
48. Michaelis J, Michaelis H, Glück E, Koller S. Prospective study of suspected associations between certain drugs administered during early pregnancy and congenital malformations. Teratology. 1983;27:57–64.
49. Di Renzo GC, Mattei A, Gojnic M, Gerli S. Progesterone and pregnancy. Curr Opin Obstet Gynecol. 2005;17:598–600.
50. Brancazio LR, Murtha AP, Heine RP. Prevention of recurrent preterm delivery by 17 alpha-hydroxyprogesterone caproate. N Engl J Med. 2003;349:1087–8.
51. O'Sullivan MD, Hehir MP, O'Brien YM, Morrison JJ. 17 alpha-hydroxyprogesterone caproate vehicle, castor oil, enhances the contractile effect of oxytocin in human myometrium in pregnancy. Am J Obstet Gynecol. 2010;202:453.
52. Yemini M, Borenstein R, Dreazen E, et al. Prevention of premature labor by 17 alpha-hydroxyprogesterone caproate. Am J Obstet Gynecol. 1985;151:574–7.
53. Meis PJ, Klebanoff M, Thom E, et al. National Institute of Child Health and Human Development Maternal-Fetal Medicine Units Network. Prevention of recurrent preterm delivery by 17 alpha-hydroxyprogesterone caproate. N Engl J Med. 2003;348:2379–85.
54. Keirse MJNC. Progestogen administration in pregnancy may prevent preterm delivery. Br J Obstet Gynecol. 1990;97:149.
55. Queisser-Luft A. Dydrogesterone use during pregnancy: overview of birth defects reported since 1977. Early Hum Dev. 1977;85:375–7.
56. Tongsong T, Srisomboon J, Wanapirak C, Sirichotiyakul S, Pongsatha S, Polsrisuthikul T. Pregnancy outcome of threatened abortion with demonstrable fetal cardiac activity: a cohort study. J Obstet Gynaecol (Tokyo 1995). 1995;21:331–5.
57. Tannirandorn Y, Sangsawang S, Manotaya S, Uerpairojkit B, Samritpradit P, Charoenvidhya D. Fetal loss in threatened abortion after embryonic/fetal heart activity. Int J Gynaecol Obstet. 2003;81:263–6.
58. Falco P, Milano V, Pilu G, David C, Grisolia G, Rizzo N, Bovicelli L. Sonography of pregnancies with first-trimester bleeding and a viable embryo: a study of prognostic indicators by logistic regression analysis. Ultrasound Obstet Gynecol. 1996;7:165–9.

Chapter 5
Progestogens and Recurrent Miscarriage

Luciano G. Nardo and Spyridon Chouliaras

1 Introduction

Progesterone has been implicated as being essential for successful embryo implantation, and for the prevention of miscarriage. In fact, progesterone was one of the first pharmacological agents to be prescribed for managing recurrent miscarriage (RM) as early as 1950 [1]. The role of progesterone for the treatment of threatened miscarriage is covered elsewhere in this book and will not be discussed here. In the current chapter we will attempt to discuss the available evidence and review the rationale for the use of progestational agents in cases of recurrent miscarriage alone.

Recurrent miscarriage (RM) affects approximately 1 % of couples trying to conceive [2]. RM has been defined as three or more consecutive spontaneous pregnancy losses with the same biological father. However, in some parts of the world including the US, RM is defined as two or more consecutive losses [3]. The fact that RM can be defined as two or more losses, complicates the interpretation of epidemiological studies and makes the available research on this subject heterogeneous. Miscarriage is defined in North America as pregnancy loss prior to 20 weeks. However, in Europe, the term miscarriage includes all pregnancy losses from the time of implantation until 24 weeks of gestation, although advances in neonatal care have resulted in babies surviving before this gestation. Primary RM refers to women with no prior successful pregnancy, while secondary RM refers to recurrent losses following a live birth.

L.G. Nardo, M.D., M.R.C.O.G. (✉)
Reproductive Health Group, Centre for Reproductive Health,
Daresbury Park, Daresbury WA4 4GE, Cheshire, UK
e-mail: lnardo@reproductivehealthgroup.co.uk

S. Chouliaras, M.D., M.R.C.O.G.
Centre for Reproductive Medicine, Kenton & Lucas Wing, St Bartholomew's Hospital,
London EC1A 7BE, UK

H.J.A. Carp (ed.), *Progestogens in Obstetrics and Gynecology*,
DOI 10.1007/978-3-319-14385-9_5

The difference between RM and sporadic miscarriage is that statistically as the demonstrated effect is being repeatedly observed the chances are that the causality is due to a systemic and recurring factor, most likely maternal in nature. Women with recurrent miscarriages tend to lose genetically normal pregnancies compared to women with sporadic miscarriage [4]. The observed incidence of recurrent miscarriage is higher than would be expected by chance alone (0.34 %) [5]. This in itself indicates that RM is a separate clinical entity to sporadic miscarriage. However, the 0.34 % chance was calculated assuming a sporadic miscarriage rate of 15 %, which is probably an underestimation.

The etiology of RM has been extensively researched. The underlying causes could be either maternal or fetal in origin. Maternal causes may include uterine factors such as endometrial pathology or congenital uterine anomalies, hormonal imbalance and insufficiency, infections, defective immunoregulation, hereditary thrombophilia, and antiphospholipid syndrome. Fetal causes include chromosomal abnormalities and structural malformations both of which are incompatible with life. Nevertheless in more than 50 % of cases no cause can be identified [6], which adds to the distress of the patient.

The prognosis of RM has been reported to be better in secondary than in primary recurrent miscarriage [7]. A descriptive cohort study of 987 women who presented with RM in a tertiary centre showed that approximately two-thirds of women with RM succeeded in having a live birth within 5 years after the first consultation, but a full third did not. There was a significantly decreased chance of at least one subsequent live birth with increasing maternal age and increasing number of miscarriages at first consultation [8].

Recurrent miscarriage is a devastating experience for the patient and a dilemma for the clinician. The emotional and psychological implications for the patient are very significant. There may be feelings of desperation, frustration, guilt, depression, low self-esteem and distrust which may overwhelm the patient. As a result the physicians have never stopped seeking ways of treating RM, especially as these patients are willing to try anything to achieve a live birth.

2 The Role of Progesterone

Progesterone was one of the first female sex hormones discovered. Its role in the successful implantation of an embryo led to progesterone being called ‘the hormone of pregnancy’. It was shown over 40 years ago that surgical removal of the corpus luteum before the eighth week of pregnancy lead to spontaneous miscarriage [9]. All progestogens are placentotrophic and their use was thought to improve trophoblastic proliferation into the spiral arteries. Progestogens are generally thought to be safe, and have become standard treatment for luteal support in assisted reproduction. Worldwide the demand from women for any treatment that provides hope drives clinicians to prescribe progestogens or progesterone. Progestogens have been used for more than half a century for the prevention of RM. Since those early years the use of progestogen preparations has been controversial, many authors have disputed the acclaimed beneficial role [10], and claimed a lack of robust evidence. The debate regarding the efficacy of this treatment is ongoing.

In this chapter we will attempt to present the pathophysiology behind the potential action of progestogens for prevention of RM as well as the best available evidence. The focus will be recurrent miscarriage of unexplained aetiology (also known as idiopathic recurrent miscarriage).

2.1 *Potential Mechanisms of Action of Progestogens in Preventing RM*

Implantation has been described as a three-stage process. The first stage comprises the apposition of the blastocyst to the endometrium. The second stage involves the adhesion of the embryo to the endometrial epithelium. In this stage the blastocyst can no longer be removed by just being flushed out. The adhesion of the blastocyst on the endometrium is due to cell surface glycoproteins, the specific mechanisms of which, are still largely unknown. The two first stages are mainly mediated by integrins, mucins, trophinins and tastins. The third stage is the invasion and embedding of the trophoblast. The invasive stages consist of two phases: early and deep invasion. Early invasion is an interplay between matrix metalloproteases (MMP) secreted by the embryo and tissue inhibitors of these proteases (TIMMP) secreted by the endometrium. During deep invasion there is an interaction between T-helper (Th) 1 cytokines preventing implantation and Th2 cytokines enhancing implantation [11].

Implantation failure has been thought to be the cause of many cases of recurrent miscarriage, as well as other reproductive sequelae such as recurrent assisted conception failure and pre-eclamptic toxaemia of pregnancy [12]. There are various ways in which progesterone influences and regulates the physiological process of implantation. This is irrespective of the mode of administration, which could be rectal, vaginal or intramuscular.

2.2 *Effects of Progesterone on the Uterus and the Endometrial Environment*

2.2.1 Endometrial Development and Luteal Phase Deficiency

Oestrogens and progesterone secreted cyclically by the ovaries, control morphological and functional changes in the endometrium, which leading to a favourable endometrial environment for implantation. This favourable environment is commonly referred to as the 'implantation or nidation window' a period where endometrial receptivity is optimal [13–15]. The nidation window is approximately 6 days after the LH surge [16] (approximately day 20 in a 28 day cycle). However, there is no consensus as to the duration of such a window [17]. Work by Navot in human embryos suggested that the implantation window lasts 4–5 days synchronous with peak progesterone concentrations [18–21].

Luteal phase deficiency or luteal phase defect (LPD) is a term that was introduced in 1949 by Georgianna-Segar Jones [22]. It has been used to describe a decrease in the amount or the duration of progesterone secretion from the corpus luteum or lack of an adequate endometrial response to ovarian steroids [22–24]. The gold standard for the diagnosis of LPD has traditionally been the morphological examination of a precisely timed luteal phase endometrial sample according to the Noyes' criteria [25]. The possible presence of two types of LPD has also been suggested: one related to immature follicles, and one where the follicles are mature [26]. Regardless of the type, progesterone supplementation aids the creation of a more receptive endometrial environment. Over the years, there have been several reports of significant variability in the histological evaluation of human endometrium dating [27, 28]. The need and usefulness of histological dating of the endometrium itself has been questioned [29], and although it has served its cause, it is nowadays considered outdated.

In recent years scientists have been seeking novel approaches to characterise the endometrium. They have utilised the ability to describe morphological changes using scanning electron microscopy (SEM), and the development of techniques that focus on molecular aspects of endometrial development. Using SEM on the uterine epithelium, both Martel et al. [30], and Nikasand et al. [31] have shown the existence of specialised cell surface structures called pinopodes (oruterodomes). The development of pinopodes has been associated with the adhesion of blastocysts to the luminal epithelium [32] and have thus been considered as markers of receptivity. Progesterone stimulates the appearance of pinopodes, whereas oestrogens cause their regression [30]. Supraphysiological levels of oestradiol such as those achieved during controlled ovarian stimulation have been associated with impairment of uterine receptivity [33]. However, other studies have failed to demonstrate a reliable pattern of pinopod expression [34, 35], and their significance as markers of endometrial receptivity remains a matter of much debate.

Despite these uncertainties, the role of progesterone in successful implantation is not disputed. The failure to synchronize the complex mechanisms involved in the crosstalk between the endometrium and the embryo results in failure of implantation. Of the numerous parameters of this synchronization process, the role of steroid hormones has been extensively studied. It is generally accepted that progesterone deficiency could contribute to the pathophysiology of recurrent pregnancy loss by delaying endometrial development. Low progesterone levels have been found in recurrent miscarriage with delayed endometrial ripening compared to normal endometrial ripening [36–38].

An increase in the secretion of oestradiol precedes ovulation and promotes the proliferation and differentiation of uterine epithelial cells. It is then followed by the secretion of progesterone, which induces stromal cell development [39].

Progesterone acts on the endometrium via specific progesterone receptors (PR) or by changing the isoforms ratio and possibly their expression level. Receptor synthesis is controlled by oestrogens through oestrogen receptors during the proliferative phase. By down-regulating oestrogen receptors, progesterone leads to a fall of both oestrogen and progesterone receptors [40].

Polymorphisms of progesterone receptors (PROGINS) have been reported to act as a risk-modulating factor in women with RM. Receptor polymorphism may cause an alteration in the biological function of PR and can be associated with an individual susceptibility to pregnancy loss, though this concept has not been confirmed in a recent meta-analysis [41, 42]. It appears that inappropriate endometrial development can occur even with sufficient progesterone levels [43], possibly due to genetic variation of progesterone receptors. The concept of absolute or relative progesterone deficiency in the pathophysiology of recurrent miscarriage, could explain why progesterone treatment may benefit some but not all women with unexplained RM.

2.2.2 Induction of Uterine Quiescence

In animal models, progesterone has been recognised as one of the major causes of inhibition of myometrial contractility. Withdrawal of progesterone is responsible for the initiation of labour. In humans, there is no detectable progesterone withdrawal, but biochemical events suggest 'functional progesterone withdrawal'. Potential mechanisms include changes in receptor isoforms and decreased myometrial sensitivity to progesterone. Nitric oxide (NO) generated in the pregnant uterus has been shown to maintain uterine relaxation [44]. Several studies have shown that progesterone enhances NO production in the endometrium [45–48]. Pharmacological withdrawal of progesterone by administration of 3-beta-hydroxysteroid dehydrogenase inhibitors or mifepristone (RU486) is associated with the onset of labour, and has been widely used to terminate early pregnancy by competitively blocking the PR [49].

2.3 Immunological Role of Progestogens

The survival of the semi-allogeneic fetus has been described as an immunological paradox. Maternal recognition of fetal antigens does not appear to compromise pregnancy. On the contrary it induces functional modifications that allow the conceptus to survive and develop. Maternal immunetolerance is established in the decidua, in a specific area known as the feto-maternal interface. It is recognised and accepted that progesterone exhibits immunomodulating properties. Indeed, there is increasing evidence that progesterone facilitates an immune environment conducive to the early development of pregnancy.

2.3.1 Involvement of Progesterone in Maternal Cytokine Production

Progesterone-induced blocking factor (PIBF) is a protein synthesised by lymphocytes of pregnant women in the presence of progesterone. Progesterone receptors on lymphocytes are moderated by the immunological recognition of pregnancy [50]. PIBF is associated with both the immunomodulatory [51] and anti-abortive [52–54] properties of progesterone. Lymphocytes from healthy pregnant women produce significantly more PIBF than those of women with pathological pregnancies.

In pregnancy there is a physiological shift in the decidual cytokine pattern from a Th1 response to a Th2 response. The cytokine shift may be modulated by PIBF [55]. Th1-type pro-inflammatory cytokines (TFN-α, IFN-γ, IL-2) support allograft rejection and are thought to be detrimental to pregnancy. TNF-, α activates natural killer (NK) cells, promotes apoptosis of the trophoblast and initiates coagulation, at least in mice [56]. Inteferon-γ can induce expression of major histocompatibility antigens on the trophoblast, where they are not normally expressed. Th2- cytokines (TGF-β2, IL-3, IL-4, IL-5, IL-10) inhibit pro-inflammatory Th1 responses, and seem to benefit pregnancy maintenance [57]. TGF-β2 induces trophoblast proliferation, IL-4 and IL-10 inhibit prothrombinase.

The activation of peripheral blood mononuclear cells (PBMC) by trophoblast antigens confirmed that women with RM have a Th1-type cytokine profile [58]. Increased production by PBMC of Th1-type cytokines and decreased levels of Th2-type cytokines have been demonstrated in non-pregnant women with recurrent early pregnancy losses [59]. Bates et al. [56] failed to demonstrate the proposed defect in the shift fromTh1 to Th2 cytokines in women with RM. Instead, increased production of IL-4 and IL-10 was shown in such women, along with reduced IFN-γ in pregnant women [56]. Nevertheless the case for a possible association between maternal Th1 dominance and recurrent miscarriage is strong. Researchers are therefore faced with the challenge of determining the optimal Th1/Th2 cytokine balance and trying to manipulate it towards an immune favourable environment.

Vaginal micronised progesterone has been reported to be associated with a decrease in IFNγ (Th-1) and increase in IL-10 (Th-2) in endocervical fluid [60]. In addition, progesterone up-regulates Leukemia inhibitory factor (LIF) mRNA expression in vitro [61]. LIF is essential for implantation in muridae.

Dydrogesterone is the most commonly used progestogen to support early pregnancy. Dydrogesterone can be administered orally and has high affinity with the PR, resembling endogenous progesterone in its pharmacology and biochemistry. Raghupathy et al., investigated the effects of dydrogesterone therapy on Th1 and Th2 cytokine production in RM. Down-regulation of Th1 cytokines (TNFα and IFNγ) and stimulation of Th2 cytokines (IL-4 and IL-6), and induction of PIBF production was reported [62]. Other researchers have suggested that the induction of PIBF production in humans could be the indirect mechanism by which dydrogesterone improves pregnancy outcome [63].

2.3.2 Involvement of Progesterone in Maternal Natural Killer (NK) Cells

There has been a considerable amount of research into the association between NK cells and recurrent miscarriage. The number of peripheral NK cells (pNK) have been associated with RM [64, 65]. However, the number or killing activity of pNK cells may not reflect the condition in the endometrium where implantation occurs; hence, the potential role of pNK cells in the pathophysiology of miscarriage remains uncertain. However, large numbers of uterine NKcells (uNK) cells may be associated with RM, but the mechanism is still unclear. High numbers of uNK cells have been found in luteal-phase endometrial biopsies of women with recurrent miscarriage compared to the pre-implantation endometrium in normal pregnancies [66]. Interestingly a recent meta-analysis that evaluated uNK cells expressed as percentage of the stromal cells failed to demonstrate a significant difference between these two groups [67]. However, the studies included in the meta-analytical pooling appeared to suffer from significant heterogeneity.

Natural killer cells, which are present in the endometrium in the luteal phase of the menstrual cycle become decidual NK cells in pregnancy. They increase in number in the first trimester [68] and are thought to control trophoblastic invasion through the production of immunoregulatory cytokines and angiogenic factors [69, 70]. Although decidual NK cells do not express progesterone receptors both the number and function of these cells are influenced indirectly by progesterone [71]. Szekeres-Bartho et al., have demonstrated that a low proportion of PIBF-positive lymphocytes is inversely related to NK cell activity and pregnancy loss [72]. PIBF is thought to contribute to the suppression of decidual NKcells cytolytic activity [73].

The role of progesterone on maternal NK cells is still being evaluated. When interpreting studies on NK cells, one should take account of the compartment (peripheral blood, endometrial, decidua etc.) in which the cells or other prognostic markers are investigated.

3 The Evidence for Progesterone Use in Recurrent Miscarriage

There have been several meta-analyses evaluating the use of progesterone for the prevention of subsequent miscarriage after RM. A recent meta-analysis in the Cochrane database showed a statistically significant reduction in miscarriage in favour of those patients randomized to the progestogen group (OR 0.39; 95 % CI 0.21–0.72) [74]. A similar outcome was reported by the same authors 2008 [75] and by two earlier meta-analyses in 2003 [76] and 1989 [36], respectively. The results of these metaanalyses do not differ as all the metaanalyses summarise the same three studies [77–79], which are now more than 50 years old and suffer from inadequate design and methodology. There have also been issues in the meta-analytical pooling. The three papers included in the earlier metaanalyses, did not account for the confounding factor of embryonic aneuploidy, nor did they stratify for the number of miscarriages. Additionally, three different progestogens were analysed, medroxyprogesterone acetate, 17-hydroxyprogesterone acetate and an implant. One study of dydrogesterone [80] was excluded from the earlier metaanalyses due to quasi-randomization rather than true randomization. However, El-Zibdeh et al. [80] is the largest to date including 180 women younger than 35 years and with a history of three or more consecutive idiopathic miscarriages in their trial. Women were randomized according to the day of the week in which they presented to receive either dydrogesterone (10 mg orally twice daily), human chorionic gonadotropin (hCG, 5,000 IU intramuscularly once every 4 days) or no treatment until week-12 of gestation. Miscarriages were significantly less common in the dydrogesterone group (13.4 %) than in the control group (29 %). Furthermore, there were no differences across the groups in respect to pregnancy complications or congenital anomalies. El-Zibdeh et al.'s [80] study, was included in the most recent meta-analysis [69], although neither placebo-controlled nor blinded. Three [77–80] of the included studies appeared to have had selection biases, and apparently only two of these studies [79, 80] enrolled women who had suffered three or more consecutive miscarriages. Two other studies [77, 78] recruited women with two or more miscarriages and provided separate pregnancy outcome data by the number of previous consecutive pregnancy losses. There was also significant variability with regard to the age of the subjects, previous successful pregnancy outcome, the investigations used to exclude other causes of miscarriage, as well as to the duration, dosage, time of initiation of treatment and choice of progestogen regimes. At present, a large randomised trial is underway (PROMISE) to assess micronised progesterone in RM. The results are eagerly awaited.

4 Safety of Progestogen Treatment for Prevention of Recurrent Miscarriage

4.1 Safety of the Mother

Worldwide millions of women have used progesterone or progestogens, and 39 million women have been assumed to have been exposed to dydrogesterone alone [81]. Few studies have accounted on the maternal adverse effects of progesterone use to prevent miscarriage [82]. However, few side effects have been reported. Typical maternal side effects of progestogens such as nausea, bloating, dizziness, breast tenderness, mood changes, cephalalgia may occur, but can also be attributed to the physiological changes occurring in early pregnancy.

Theoretically there could be concerns that progesterone may delay spontaneous miscarriage, by promotion of uterine quiescence or even aid retention of chromosomally abnormal embryos. However, it seems that the mechanism in immunologically mediated abortion may be different to that in abortion due to aneuploidy. In the latter event implantation fails altogether, whereas in immunologically mediated abortion there is adequate implantation and subsequently an immunologically mediated process of vasculitis, inflammation leading to thrombus formation. Progestogens only influence chromosomally normal embryos by regulating immunomodulation [83, 84].

4.2 *Safety of the Fetus*

There have been reports suggesting an association between intrauterine exposure to progestogens in the first trimester of pregnancy and genital abnormalities in both male and female fetuses. Some progestogens such as ethisterone have been thought to induce mild virilisation of the external genitalia in the female fetus [85, 86]. However, virilisation has only been seen in rat fetuses. The lack of evidence in humans has made it impossible to quantify the risk. Carmichael et al. [87] reported that maternal intake of progestins in early pregnancy is associated with an increased risk of hypospadias in the male fetus due to an anti-androgenic effect (OR 3.7, 95 % CI 2.3, 6.0) [81]. Other studies do not indicate an increased risk with exposure to progestins. Interestingly dydrogesterone has less of an antiandrogenic effect than progesterone itself. The urogenital groove is fused by 16 weeks of gestation, in order to avoid the antiandrogenic effect, some authors have recommended that progesterone containing medications should be avoided in the first trimester of pregnancy [88].

5 Future Research

The lack of robust data has generated the need for well-conducted studies to assess the validity of intervention with progesterone supplementation in RM. In a small unpublished study it was reported that 90 % of obstetricians and gynaecologists called for a definitive placebo controlled randomised trial [89]. The potential pitfalls for such a study are many, and careful design is of paramount importance. Matching age and number of miscarriages is one component. Stratification for primary versus secondary miscarriage is another factor which has to be taken into account. Unfortunately heterogeneity in human populations is unavoidable even in patients with exactly the same clinical characteristics. All pregnancies should have been conceived with the same partner, which is again almost impossible to elicit. The exclusion of other causes of miscarriage is also difficult to ascertain, as there are subjective variables such as the presence of congenital uterine anomalies. One randomized study is underway and the results are eagerly awaited. The study will hopefully shed more light in the use of progestogens for RM. The 'Progesterone in recurrent miscarriage (PROMISE) study' (ISRCTN92644181) is a randomised, double blind, placebo controlled multicentre trial based in UK and the Netherlands studying the effect of progesterone treatment given in the first trimester of pregnancy in women with a history of unexplained recurrent miscarriages, who conceive spontaneously. The intervention being assessed is vaginal micronised progesterone pessaries 400 mg twice daily starting as soon after a positive pregnancy test as possible (but no later than 6 weeks of gestation) and continued up to 12 weeks of gestation. The primary outcome is live birth beyond 24 completed weeks. The results are awaited in March 2015 [89].

Conclusions

Progesterone is a 'pro-gestational' agent that maintains the pregnant state. The concept of the association of RM with retarded endometrial development in the peri-implantation period or luteal phase defect is well recognised. The immunomodulatory function of progesterone appears to be decisive in early pregnancy. It is therefore quite possible that there may be a role for its use in women with unexplained RM. The paucity of good quality evidence about the efficacy of progesterone in women with a history of recurrent early pregnancy loss is responsible for contradictory and ever-changing views amongst clinicians.

Research into RM is ongoing and focused on several areas, mostly at the bio-molecular level. Factors such as circulating pro-coagulant microparticles, glycoproteins (for example, hCG and glycodelin), leptin receptors and TNF-α inhibitors are being investigated. As our understanding of the pathophysiology of recurrent miscarriage improves new causes are likely to be identified. This will lead to individualisation of treatment for RM.

The role of progesterone treatment for unexplained recurrent miscarriage will be re-evaluated after new data from ongoing studies becomes available. However, even if these studies fail to demonstrate evidence of benefit, maybe in the future the validity of the results could be questioned as the criteria for diagnosis of unexplained RM in this high-risk population may change dramatically in the light of novel basic science research findings.

References

1. Benson RC. Habitual abortion. California Medicine. 1950;72(6):442–6.
2. Stirrat GM. Recurrent miscarriage. Lancet. 1990;336(8716):673–5.
3. Practice Committee of the American Society for Reproductive Medicine. Definitions of infertility and recurrent pregnancy loss: a committee opinion. Fertil Steril. 2013;99(1):63.
4. Ogasawara M, Aoki K, Okada S, Suzumori K. Embryonic karyotype of abortuses in relation to the number of previous miscarriages. Fertil Steril. 2000;73:300–4.
5. Regan L. Recurrent miscarriage. BMJ. 1991;302(6776):543–4.
6. Rai R, Regan L. Recurrent miscarriage. Lancet. 2006;368(9535):601–11.

7. Regan L, Braude PR, Trembath PL. Influence of past reproductive performance on risk of spontaneous abortion. BMJ. 1989;299(6698):541–5.
8. Lund M, Kamper-Jorgensen M, Nielsen HS, Lidegaard O, Andersen AM, Christiansen OB. Prognosis for live birth in women with recurrent miscarriage: what is the best measure of success? Obstet Gynecol. 2012;119(1):37–43.
9. Csapo AI, Pulkkinen MO, Wiest WG. Effects of luteectomy and progesterone replacement therapy in early pregnant patients. Am J Obstet Gynecol. 1973;115(6):759–65.
10. Progestogen therapy in habitual abortion. Br Med J. 1963;1(5326):283–4.
11. Nardo LG, Sallam HN. Progesterone supplementation to prevent recurrent miscarriage and to reduce implantation failure in assisted reproduction cycles. Reprod Biomed Online. 2006;13(1):47–57.
12. Edwards RG. Clinical approaches to increasing uterine receptivity during human implantation. Hum Reprod. 1995;10 Suppl 2:60–6.
13. Psychoyos A. Hormonal control of ovoimplantation. Vitam Horm. 1973;31:201–56.
14. Psychoyos A. Hormonal control of uterine receptivity for nidation. J Reprod Fertil Suppl. 1976;25:17–28.
15. Tabibzadeh S, Babaknia A. The signals and molecular pathways involved in implantation, a symbiotic interaction between blastocyst and endometrium involving adhesion and tissue invasion. Hum Reprod. 1995;10(6):1579–602.
16. Sarani SA, Ghaffari-Novin M, Warren MA, Dockery P, Cooke ID. Morphological evidence for the 'implantation window' in human luminal endometrium. Hum Reprod. 1999;14(12):3101–6.
17. Gruidl M, Buyuksal A, Babaknia A, Fazleabas AT, Sivarajah S, Satyaswaroop PG, et al. The progressive rise in the expression of alpha crystallin B chain in human endometrium is initiated during the implantation window: modulation of gene expression by steroid hormones. Mol Hum Reprod. 1997;3(4):333–42.
18. Laufer N, Navot D, Schenker JG. The pattern of luteal phase plasma progesterone and estradiol in fertile cycles. Am J Obstet Gynecol. 1982;143(7):808–13.
19. Navot D, Anderson TL, Droesch K, Scott RT, Kreiner D, Rosenwaks Z. Hormonal manipulation of endometrial maturation. J Clin Endocrinol Metab. 1989;68(4):801–7.
20. Navot D, Bergh PA, Williams M, Garrisi GJ, Guzman I, Sandler B, et al. An insight into early reproductive processes through the in vivo model of ovum donation. J Clin Endocrinol Metab. 1991;72(2):408–14.
21. Navot D, Scott RT, Droesch K, Veeck LL, Liu HC, Rosenwaks Z. The window of embryo transfer and the efficiency of human conception in vitro. Fertil Steril. 1991;55(1):114–8.
22. Jones G. Some newer aspects of the management of infertility. J Am Med Assoc. 1949;141(16):1123–9.
23. Dawood MY. Corpus luteal insufficiency. Curr Opin Obstet Gynecol. 1994;6(2):121–7.
24. Cumming DC, Honore LH, Scott JZ, Williams KP. The late luteal phase in infertile women: comparison of simultaneous endometrial biopsy and progesterone levels. Fertil Steril. 1985;43(5):715–9.
25. Noyes RWHA, Rock J. Dating the endometrial biopsy. Fertil Steril. 1950;1:3–25.
26. Check JH. Progesterone therapy versus follicle maturing drugs—possible opposite effects on embryo implantation. Clin Exp Obstet Gynecol. 2002;29(1):5–10.
27. Li TC, Dockery P, Rogers AW, Cooke ID. How precise is histologic dating of endometrium using the standard dating criteria? Fertil Steril. 1989;51(5):759–63.
28. Myers ER, Silva S, Barnhart K, Groben PA, Richardson MS, Robboy SJ, et al. Interobserver and intraobserver variability in the histological dating of the endometrium in fertile and infertile women. Fertil Steril. 2004;82(5):1278–82.
29. Coutifaris C, Myers ER, Guzick DS, Diamond MP, Carson SA, Legro RS, et al. Histological dating of timed endometrial biopsy tissue is not related to fertility status. Fertil Steril. 2004;82(5):1264–72.
30. Martel D, Monier MN, Roche D, Psychoyos A. Hormonal dependence of pinopode formation at the uterine luminal surface. Human Reprod. 1991;6(4):597–603.

31. Nikas G, Drakakis P, Loutradis D, Mara-Skoufari C, Koumantakis E, Michalas S, et al. Uterine pinopodes as markers of the 'nidation window' in cycling women receiving exogenous oestradiol and progesterone. Human Reprod. 1995;10(5):1208–13.
32. Psychoyos A. Uterine receptivity for nidation. Ann N Y Acad Sci. 1986;476:36–42.
33. Nardo LG, Sabatini L, Rai R, Nardo F. Pinopode expression during human implantation. Eur J Obstet Gynecol Reprod Biol. 2002;101(2):104–8.
34. Quinn C, Ryan E, Claessens EA, Greenblatt E, Hawrylyshyn P, Cruickshank B, et al. The presence of pinopodes in the human endometrium does not delineate the implantation window. Fertil Steril. 2007;87(5):1015–21.
35. Acosta AA, Elberger L, Borghi M, Calamera JC, Chemes H, Doncel GF, et al. Endometrial dating and determination of the window of implantation in healthy fertile women. Fertil Steril. 2000;73(4):788–98.
36. Daya S. Efficacy of progesterone support for pregnancy in women with recurrent miscarriage. A meta-analysis of controlled trials. Br J Obstet Gynaecol. 1989;96(3):275–80.
37. Babalioglu R, Varol FG, Ilhan R, Yalcin O, Cizmecioglu F. Progesterone profiles in luteal-phase defects associated with recurrent spontaneous abortions. J Assist Reprod Genet. 1996;13(4):306–9.
38. Li TC, Spuijbroek MD, Tuckerman E, Anstie B, Loxley M, Laird S. Endocrinological and endometrial factors in recurrent miscarriage. BJOG. 2000;107(12):1471–9.
39. Norwitz ER, Schust DJ, Fisher SJ. Implantation and the survival of early pregnancy. N Engl J Med. 2001;345(19):1400–8.
40. Bergeron C. Morphological changes and protein secretion induced by progesterone in the endometrium during the luteal phase in preparation for nidation. Human Reprod. 2000;15 Suppl 1:119–28.
41. Su MT, Lee IW, Chen YC, Kuo PL. Association of progesterone receptor polymorphism with idiopathic recurrent pregnancy loss in Taiwanese Han population. J Assist Reprod Genet. 2011;28(3):239–43.
42. Su M-T, Lin S-H, Chen Y-C. Association of sex hormone receptor gene polymorphisms with recurrent pregnancy loss: a systematic review and meta-analysis. Fertil Steril. 2011;96(6): 1435–44.e1.
43. Li TC, Tuckerman EM, Laird SM. Endometrial factors in recurrent miscarriage. Hum Reprod Update. 2002;8(1):43–52.
44. Dong YL, Gangula PR, Yallampalli C. Nitric oxide synthase isoforms in the rat uterus: differential regulation during pregnancy and labour. J Reprod Fertil. 1996;107(2):249–54.
45. Simoncini T, Caruso A, Giretti MS, Scorticati C, Fu X-D, Garibaldi S, et al. Effects of dydrogesterone and of its stable metabolite, 20-α-dihydrodydrogesterone, on nitric oxide synthesis in human endothelial cells. Fertil Steril. 2006;86(4 Suppl):1235–42.
46. Potdar N, Konje JC. The endocrinological basis of recurrent miscarriages. Curr Opin Obstet Gynecol. 2005;17(4):424–8.
47. Arrowsmith S, Kendrick A, Wray S. Drugs acting on the pregnant uterus. Obstet Gynaecol Reprod Med. 2010;20(8):241–7.
48. Andronowska A, Chruściel M. Influence of estradiol-17β and progesterone on nitric oxide (NO) production in the porcine endometrium during first half of pregnancy. Reprod Biol. 2008;8(1):43–55.
49. Couzinet B, Le Strat N, Ulmann A, Baulieu EE, Schaison G. Termination of early pregnancy by the progesterone antagonist RU 486 (Mifepristone). N Engl J Med. 1986;315(25): 1565–70.
50. Szekeres-Bartho J, Szekeres G, Debre P, Autran B, Chaouat G. Reactivity of lymphocytes to a progesterone receptor-specific monoclonal antibody. Cell Immunol. 1990;125(2):273–83.
51. Szekeres-Bartho J, Reznikoff-Etievant MF, Varga P, Pichon MF, Varga Z, Chaouat G. Lymphocytic progesterone receptors in normal and pathological human pregnancy. J Reprod Immunol. 1989;16(3):239–47.
52. Szekeres-Bartho J, Par G, Dombay G, Smart YC, Volgyi Z. The antiabortive effect of progesterone-induced blocking factor in mice is manifested by modulating NK activity. Cell Immunol. 1997;177(2):194–9.

53. Szekeres-Bartho J, Par G, Szereday L, Smart CY, Achatz I. Progesterone and non-specific immunologic mechanisms in pregnancy. Am J Reprod Immunol. 1997;38(3):176–82.
54. Arck PC, Rücke M, Rose M, Szekeres-Bartho J, Douglas AJ, Pritsch M, et al. Early risk factors for miscarriage: a prospective cohort study in pregnant women. Reprod Biomed Online. 2008;17(1):101–13.
55. Szekeres-Bartho J, Faust Z, Varga P, Szereday L, Kelemen K. The immunological pregnancy protective effect of progesterone is manifested via controlling cytokine production. Am J Reprod Immunol. 1996;35(4):348–51.
56. Bates MD, Quenby S, Takakuwa K, Johnson PM, Vince GS. Aberrant cytokine production by peripheral blood mononuclear cells in recurrent pregnancy loss? Human Reprod. 2002;17(9):2439–44.
57. Wegmann TG, Lin H, Guilbert L, Mosmann TR. Bidirectional cytokine interactions in the maternal-fetal relationship: is successful pregnancy a TH2 phenomenon? Immunol Today. 1993;14(7):353–6.
58. Raghupathy R, Makhseed M, Azizieh F, Omu A, Gupta M, Farhat R. Cytokine production by maternal lymphocytes during normal human pregnancy and in unexplained recurrent spontaneous abortion. Human Reprod. 2000;15(3):713–8.
59. Hill JA. T-helper 1-type immunity to trophoblast: evidence for a new immunological mechanism for recurrent abortion in women. Human Reprod. 1995;10 Suppl 2:114–20.
60. Alimohamadi S, Javadian P, Gharedaghi MH, Javadian N, Alinia H, Khazardoust S, et al. Progesterone and threatened abortion: a randomized clinical trial on endocervical cytokine concentrations. J Reprod Immunol. 2013;98(1–2):52–60.
61. Aisemberg J, Vercelli CA, Bariani MV, Billi SC, Wolfson ML, Franchi AM. Progesterone is essential for protecting against LPS-induced pregnancy loss. LIF as a potential mediator of the anti-inflammatory effect of progesterone. PLoS One. 2013;8(2):e56161.
62. Raghupathy R, Al Mutawa E, Makhseed M, Azizieh F, Szekeres-Bartho J. Modulation of cytokine production by dydrogesterone in lymphocytes from women with recurrent miscarriage. BJOG. 2005;112(8):1096–101.
63. Kalinka J, Szekeres-Bartho J. The impact of dydrogesterone supplementation on hormonal profile and progesterone-induced blocking factor concentrations in women with threatened abortion. Am J Reprod Immunol. 2005;53(4):166–71.
64. Kwak JY, Beer AE, Kim SH, et al. Immunopathology of the implantation site utilizing monoclonal antibodies to natural killer cells in women with recurrent pregnancy losses. Am J Reprod Immunol. 1999;41:91–8.
65. Aoki K, Kajiura S, Matsumoto Y, et al. Preconceptual natural killer cell activity as a predictor of miscarriage. Lancet. 1995;345:1340–2.
66. Clifford K, Flanagan AM, Regan L. Endometrial CD56+ natural killer cells in women with recurrent miscarriage: a histomorphometric study. Human Reprod. 1999;14(11):2727–30.
67. Seshadri S, Sunkara SK. Natural killer cells in female infertility and recurrent miscarriage: a systematic review and meta-analysis. Hum Reprod Update. 2013;20(3):429–38.
68. King A, Burrows T, Verma S, Hiby S, Loke YW. Human uterine lymphocytes. Hum Reprod Update. 1998;4(5):480–5.
69. Hanna J, Goldman-Wohl D, Hamani Y, Avraham I, Greenfield C, Natanson-Yaron S, et al. Decidual NK cells regulate key developmental processes at the human fetal-maternal interface. Nat Med. 2006;12(9):1065–74.
70. Sargent IL, Borzychowski AM, Redman CW. Immunoregulation in normal pregnancy and pre-eclampsia: an overview. Reprod Biomed Online. 2006;13(5):680–6.
71. Henderson TA, Saunders PT, Moffett-King A, Groome NP, Critchley HO. Steroid receptor expression in uterine natural killer cells. J Clin Endocrinol Metab. 2003;88(1):440–9.
72. Szekeres-Bartho J, Faust Z, Varga P. The expression of a progesterone-induced immunomodulatory protein in pregnancy lymphocytes. Am J Reprod Immunol. 1995;34(6):342–8.
73. Faust Z, Laskarin G, Rukavina D, Szekeres-Bartho J. Progesterone-induced blocking factor inhibits degranulation of natural killer cells. Am J Reprod Immunol. 1999;42(2):71–5.
74. Haas DM, Ramsey PS. Progestogen for preventing miscarriage. Cochrane Database Syst Rev. 2013;10:Cd003511.

75. Haas DM, Ramsey PS. Progestogen for preventing miscarriage. Cochrane Database Syst Rev. 2008;2:Cd003511.
76. Oates-Whitehead RM, Haas DM, Carrier JA. Progestogen for preventing miscarriage. Cochrane Database Syst Rev. 2003;4:Cd003511.
77. Swyer GI, Daley D. Progesterone implantation in habitual abortion. Br Med J. 1953;1(4819):1073–7.
78. Goldzieher JW. Double-blind trial of a progestin in habitual abortion. JAMA. 1964;188:651–4.
79. Levine L. Habitual abortion. A controlled study of progestational therapy. West J Surg Obstet Gynecol. 1964;72:30–6.
80. El-Zibdeh MY. Dydrogesterone in the reduction of recurrent spontaneous abortion. J Steroid Biochem Mol Biol. 2005;97(5):431–4.
81. Queisser-Luft A. Dydrogesterone use during pregnancy: overview of birth defects reported since 1977. Early Hum Dev. 2009;85:375–7.
82. Moller KJ, Bojsen-Moller B, Fuchs F, Villumsen A. Treatment of threatened abortion with 6-methyl-17-acetoxyprogesterone. Preliminary report on a double-blind controlled study. Acta Obstet Gynecol Scand. 1964;42 Suppl 6:124–5.
83. Quack KC, Vassiliadou N, Pudney J, Anderson DJ, Hill JA. Leukocyte activation in the decidua of chromosomally normal and abnormal fetuses from women with recurrent abortion. Human Reprod. 2001;16(5):949–55.
84. Raghupathy R. Th1-type immunity is incompatible with successful pregnancy. Immunol Today. 1997;18(10):478–82.
85. Whalen RE, Peck CK, LoPiccolo J. Virilization of female rats by prenatally administered progestin. Endocrinology. 1966;78(5):965–70.
86. Becker KL, Rebar RW. Principles and practice of endocrinology and metabolism. Philadelphia: Lippincott Williams & Wilkins Publishers; 2002.
87. Carmichael SL, Shaw GM, Laurent C, Croughan MS, Olney RS, Lammer EJ. Maternal progestin intake and risk of hypospadias. Arch Pediatr Adolesc Med. 2005;159(10):957–62.
88. Briggs GG, Freeman RK, Yaffe SJ. Drugs in pregnancy and lactation. 9th ed. Philadelphia: Lippincott Williams & Wilkins; 2009.
89. Coomarasamy A, Truchanowicz EG, Rai R. Does first trimester progesterone prophylaxis increase the live birth rate in women with unexplained recurrent miscarriages? BMJ. 2011;342:d1914.

Chapter 6
Progestogens in Preterm Labour

Gian Carlo Di Renzo, Irene Giardina, Giulia Babucci, Chiara Antonelli, Sandro Gerli, and Graziano Clerici

1 Introduction

Pre-term births (PTB) (below 37 weeks) is one of the major health challenges to-day. The incidence is high, accounting for 5–9 % of births in Europe [1]. In addition, the incidence is increasing. The incidence was 9.5 % in the U.S. in 1981, rising to 12.7 % in 2005 [2]. Perinatal mortality increases more than three times in women with PTB (51.7/1,000 births), and is a leading cause of long-term morbidity, including neurodevelopmental handicap, cerebral palsy, seizures, blindness, deafness and non-neurological disorders, such as bronchopulmonary dysplasia & retinopathy of prematurity. Additionally, initial hospital care of infants born at 25–27 weeks costs 28 times as much as for those born at term in the U.S.

The diverse etiology of preterm delivery makes prediction difficult. Previous PTB, cervical effacement or dilatation on clinical vaginal examination or shortening of the length of the cervical canal on ultrasound are predisposing factors. Similarly, the presence of contractions may indicate the start of preterm labour. In pre-term labor (PTL) various tocolytics have been used to stop contractions including: (1) β adrenergic stimulators, known to relax smooth muscle such as ritodrine, terbutalin etc. (2) Prostaglandin synthetase inhibitors such as indomethacin. (3) Calcium channel blockers such as nifedipine, and (4) oxytocin inhibitors such as atosiban. The efficacy of all of these is controversial.

G.C. Di Renzo, MD, PhD, FRCOG (hon), FACOG (hon) (✉) • I. Giardina
G. Babucci • C. Antonelli • S. Gerli • G. Clerici
Santa Maria della Misericordia University Hospital, San Sisto, Perugia 06132, Italy
e-mail: giancarlo.direnzo@unipg.it

H.J.A. Carp (ed.), *Progestogens in Obstetrics and Gynecology*,
DOI 10.1007/978-3-319-14385-9_6

A growing body of evidence suggests that progesterone plays a role in preventing labour. Progesterone decreases prostaglandin synthesis, reduces cervical stromal degradation, alters the barrier to ascending inflammation and or infection in the cervix. It reduces gap junction formation, decreases the conduction and frequency of myometrial contractions, increases the stimulation threshold of the myometrium and decreases spontaneous myometrial activity. Progesterone alters estrogen synthesis in fetal membranes and the placenta, alters fetal endocrine-mediated effects. It decreases the number of oxytocin receptors and suppresses the inflammatory cascade.

This chapter concentrates on the mechanisms whereby progestogens may prevent PTL, the optimal progestogens, and their route of administration, and the place of progestogens in current obstetric practice.

2 Role of Progesterone in Pregnancy

Progesterone is an essential hormone in the process of reproduction. It has been proposed and used in the treatment of different gynaecological pathologies, such as endometrial hyperplasia, dysfunctional uterine bleeding, amenorrhoea, luteal phase deficiency and premenstrual syndrome, as well as being used as a contraceptive tool alone or in combination with estrogens, in assisted reproductive technologies and in the maintenance of pregnancy [3]. The role of progesterone in the maintenance of pregnancy comprises: modulation of maternal immune response [4–6], suppression of inflammatory response [7], reduction of uterine contractility [8–10], and improvement of the utero-placental circulation [11, 12]. Some of the cytokine effects associated with pre-term births have been defined. Increased amniotic levels of the proinflammatory cytokines have such as IL-6, IL-8 and TNFα have been correlated with preterm births [13]. The physiological effects of progesterone are fully discussed in Chap. 1. The cytokine modulatory effects have been best studied in the synthetic progestogen dydrogesterone. Dydrogesterone inhibits IFNγ & TNFα, but increases IL-4 production [14]. Dydrogesterone induces the production of a progesterone induced blocking factor (PIBF). In Preterm labour, dydrogesterone is associated with significantly higher serum levels of IL-10, increased PIBF production & lower concentrations of IFNγ than controls [15]. Progesterone itself has been reported to up-regulate LIF mRNA expression in vitro [16].

All these effects have been used to explain the therapeutic application of progesterone during pregnancy, whether progestogens are targeted to the prevention and treatment of threatened miscarriage, recurrent miscarriage or preterm birth. Regarding the myometrial effect, progesterone has been shown to possess tocolytic action on the myometrium both in vitro and in vivo during pregnancy. It has been demonstrated, that adequate progesterone concentrations in the myometrium are able to counteract the stimulatory activity of both prostaglandins and oxytocin. Progesterone decreases the concentration of myometrial oxytocin receptors, (counteracting the effect of estrogens). Progesterone also decreases the number and activity of gap junctions [17]. Moreover, progesterone and its metabolites induce uterine

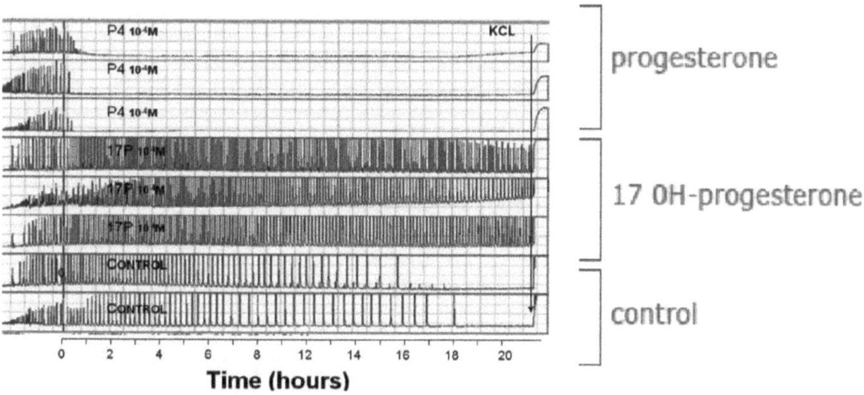

Fig. 6.1 Tocolytic effects of progesterone: P4 versus 17OHP: changes in contractility in progesterone and 17OHP treated myometrial strips. Modified from Chanrachakul B et al. Am J Obstet Gynecol 2005 and Ruddock NK et al., Am J Obstet Gynecol 2008 [10, 20]

quiescence through interactions between nuclear and membrane progesterone receptors [9, 18, 19].

The in vivo effect of progesterone has also been shown to be dose dependent. Only high-doses of progesterone are tocolytic in early pregnancy. High dosages are also effective in the maintenance of uterine quiescence during cervical cerclage (during the first trimester of pregnancy) and/or following abdominal surgery (e.g. appendectomy) [17]. However, the progestogen 17 alpha-hydroxyprogesterone caproate (17 OHP-C) does not directly inhibit human myometrial contractions in vitro. Surprisingly, in fact, 17 OHP-C has been shown to stimulate contractility in a dose dependent manner [20] (Fig. 6.1).

Since the 1990s, progesterone has been studied for its possible role in preterm birth. Keirse in his meta-analysis of placebo-controlled trials, investigated the possible prophylactic use of 17 OHP-C in women at a high risk of pre-term birth. This analysis no effect in miscarriage, but did demonstrate a reduction in the incidence of preterm birth and the incidence of low birthwheight babies [21].

3 Key Differences in the Route of Administration

Although the pharmacokinetics and pharmacodynamics of progesterone have been well studied since 1935 when progesterone was first synthesized, its use in the pathophysiology of pregnancy remains controversial. One of these concerns is the optimal route of administration [22]. The rate of absorption is dependent on which pharmaceutical form is used, the blood flow at the site of administration and the solubility in the tissues into which the drug is administered. Progesterone can be administered by many different routes, but the main are oral, vaginal and intramuscular (Table 6.1).

Table 6.1 Routes of administration of progesterone

Different routes of administration	
Different pharmacokinetics and dynamics	
Intramuscular Supraphysiological plasma concentrations	
Oral • Rapid increase n plasma concentration followed by a gradual decrease • First liver pass effect with several biological active metabolites • Specific activity on different target organs (uterus, brain…)	*Metabolism* – In the gut (bacteria with 5b-reductase activity) – In the intestinal wall (5a-reductase activity) – In the liver (5b-reductase, 3a- and 20a-hydrodylase activities)
Vaginal • Stable plasma concentrations and consistent tissue levels • First uterine pass effect with targeted delivery into the endometrium • Minimal systemic effects	*Metabolism* – Normal vaginal bacteria and mucosa seem devoid 5a- and 5b-reductases – After vaginal, only a small increase in 5a-pregnanolone observed and 5b-pregnanolone levels were not affected

3.1 Oral Administration

Oral administration is associated with better patient compliance, but has several disadvantages, the main disadvantage being extreme variability in plasma concentrations due to individual variability in gastric filling and enteropathic circulation. Food may influence the rate and extent of drug absorption by reducing the rate of gastric emptying, decreasing gastrointestinal motility, increasing gastrointestinal secretions and increasing splanchnic blood flow. Oral administration has another major shortcoming: after absorption, progesterone is first passed to the liver by the portal vein. The metabolites produced by the liver interfere with progesterone action; as shown by the discrepancy between progesterone levels and endometrial histology [23].

A number of synthetic oral progestogens have been developed to overcome progesterone's low oral bioavailability. Their pharmacologic effects, however, differ from progesterone itself, and may result in increased androgenic effects, fluid retention, alterations in high density lipoproteins, headaches, mood disturbances, and possible teratogenicity dependant on the individual progestogen. Dydrogesterone however, seems to have no androgenic or anti-androgenic effect (See Table 2 in Chap. 2). Oral progesterone may also have other side effects such as nausea, headache and sleepiness. In an attempt to improve the characteristics of oral administration, micronized progesterone has been introduced. It is manufactured from chemicals derived from plants (Mexican wild yams, *Dioscorea barbasco*), yet it has a molecular structure identical to human progesterone [24]. *Micronization of progesterone* into particle sizes of <10 μm increases the available surface area, enhances the aqueous dissolution rate and intestinal absorption of progesterone. Suspension in oil and packaging in a gelatin capsule has been shown to further enhance the intestinal absorption of micronized progesterone [25].

3.2 Vaginal Administration

The vaginal route results in higher concentrations in the uterus [17, 22]. Bulletti et al. [26] tried to verify the hypothesis of a "first uterine pass effect," suggested by the evidence of higher than expected uterine tissue concentrations after vaginal administration of progesterone. Three different hypotheses have been reported to explain the first uterine pass: direct diffusion through or between cells of the vagina to cells of the uterus, portal-like arrangements of lymphatics linking the upper vagina to the uterus, and a counter circulation system, much like a portal system, with vein to artery diffusion between the upper vagina and uterus. Bulletti et al. [26] demonstrated that a "first uterine pass effect" occurs when drug is delivered vaginally, confirming that the vaginal route permits targeted drug delivery to the uterus, maximizing the desired effects while minimizing the potential for adverse systemic effects (Fig. 6.2).

After vaginal administration of progesterone, uterine tissue concentration have been found to exceed the levels achieved by systemic administration by more than tenfold; but the plasma levels are more than seven times higher after systemic administration compared to vaginal administration. The time to peak concentration is generally slightly less than after oral administration of micronized preparation. Moreover, after vaginal administration, plasma concentrations display a plateau-like profile, with a more constant concentration over time. Peak plasma levels of progesterone seem to be variable and not consistently greater or less than corresponding peak plasma values obtained after oral administration of micronized

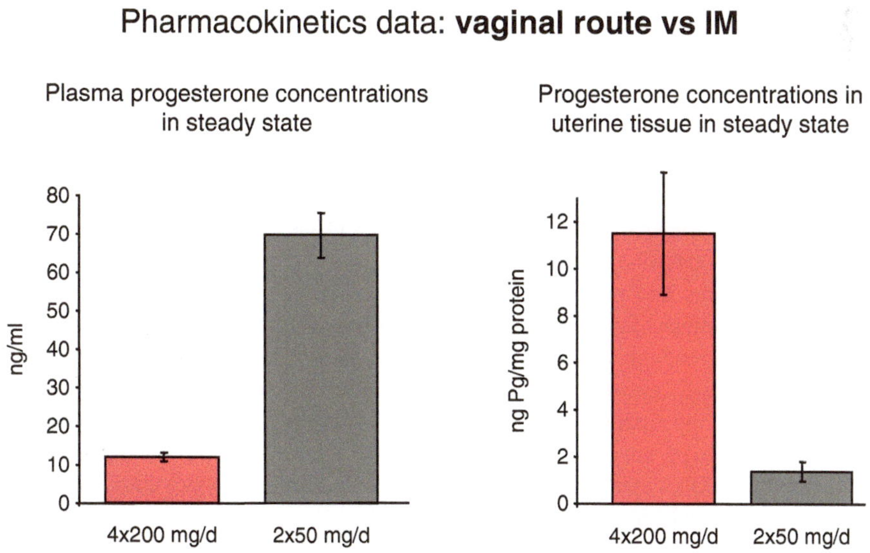

Fig. 6.2 Comparison between vaginal and intramuscular route of administration. Modified from Miles A et al. Fertil Steril 1994 [45]

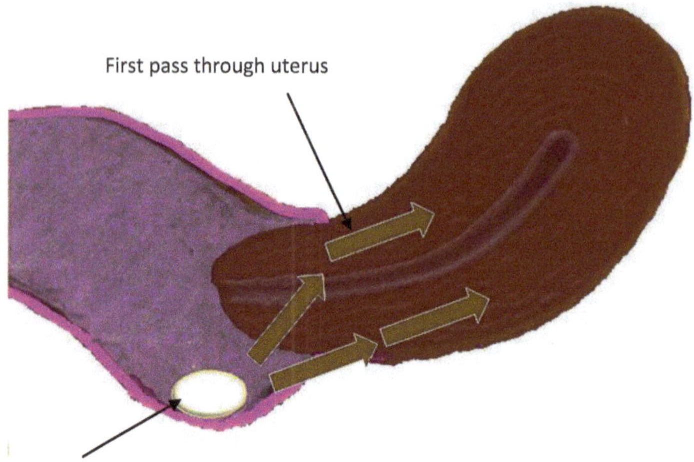

Fig. 6.3 Vaginal administration of progesterone: first uterine pass effect

progesterone [25]. Cicinelli et al. demonstrated the direct transport of progesterone from vagina to uterus comparing progesterone concentrations in serum and endometrial tissue from hysterectomy specimens after vaginal or intramuscular administration of progesterone gel. They demonstrated that the ratios of endometrial to serum progesterone concentrations were markedly higher in women who received vaginal progesterone [27] (Fig. 6.3).

3.3 Intramuscular Administration

Historically, the most common delivery method of progesterone has been *intramuscular* [25]. Progesterone administered intramuscularly can lead to local discomfort and extreme pain, and occasionally induces non-septic abscesses at the injection site. A large number of other side effects have been described including: hypersensitivity reactions, cough, dyspnea, tiredness, dizziness, genital itching, & increased risk of gestational diabetes, mood swings, headaches, bloating, abdominal pain, perineal pain, constipation, diarrhea, nausea, vomiting, joint pain, depression, decreased sex drive, nervousness, sleepiness, breast enlargement, breast pain, dysuria, polyuria, UTI, vaginal discharge, fever, flu-like symptoms, back pain, leg pain, sleep disorder, upper respiratory infection, asthma, acne and pruritus.

However, intramuscular administration is the route that results in optimal blood levels, can be administered in patient with vaginal bleeding and needs only one single daily dose [17, 22].

3.4 Other Routes of Administration

Other routes have been described. Progesterone can be administered by a transdermal patch *as* progesterone can penetrate the skin but it is rapidly metabolized by the 5-α-reductase enzyme, which converts it to 5-α-dihydro-progesterone, thereby lowering plasma progesterone levels. Rectal administration, is associated with variable absorption and there is insufficient scientific evidence concerning effects on the endometrium after rectal administration.

4 Progesterone and Preterm Birth

As stated above, PTB is the leading cause of perinatal mortality and morbidity. Its incidence has not declined (12 % of all birth) over the last 20 years, and, due to its long-term neuro-developmental sequelae, it is one of the major financial drains on health and educational resources [10, 28]. It is the major challenge in Obstetrics to-day to reduce the number of PTB's or at least prolong pregnancy until fetal maturity is sufficient to ensure survival without excess morbidity.

The mechanism of human parturition is the expression of anatomic, biochemical, physiologic and clinical events that occur in the mother and in the fetus in both term and preterm labour. This pathway comprises: decidual/fetal membrane activation, increased uterine contractility and cervical ripening (dilatation and effacement). Preterm labor is the consequence of the pathologic activation of one or more of these elements. The apparent loss of progesterone sensitivity at term could be a consequence of several different mechanisms including: alterations in progesterone receptors (PR) isoform ratios, the anti-inflammatory function of progesterone, the catabolism of progesterone in the uterus into inactive compounds, changes in cofactor protein levels affecting PR transactivation and inflammation-induced trans-repression of PR [17].

Before undertaking any therapeutic strategy, careful identification is needed, so as to detect manageable conditions and fetal and/or maternal contraindication. The real challenge is to develop sensitive and specific tests that reliably detect these pregnancy changes before they became irreversible, and to find effective interventions capable of arresting the process of preterm labor and to enhance the effectiveness of current available interventions.

Regarding the management of threatened preterm labor, tocolysis and administration of corticosteroids to induce lung maturation are the first therapeutic tools; additionally bed rest and hydration are usually recommended in the management of these patients, although none have been shown to be clearly effective. Progesterone, other related synthetic compounds such as 17 OHP-C and other progestogens, have been tested in clinical trials for the *prevention of preterm birth* [29]. The three main groups of patients considered at particularly high risk include: patients with a previous preterm birth, patients with short cervix and patients with multiple pregnancy. Studies using synthetic progestogens to reduce the incidence of preterm delivery have been reported. However, there are mixed results. Progesterone itself however, has been demonstrated to be beneficial, especially in view of cost, availability and biological safety [30, 31].

The vast majority of clinical trials have been performed with various formulations of either microized progesterone or 17 OHP-C. The first has been administered through the vaginal route and the latter by weekly intramuscular injection. The use of both micronized progesterone and 17 OHP-C has been advised in asymptomatic women with a prior history of preterm birth as early prophylaxis for preterm, birth and in single pregnant, nulliparous women with asymptomatic cervical shortening (15 mm) detected with trans-vaginal ultrasound at midgestation. The tocolytic effect of oral micronized progesterone was first reported in a 1986 report by Erny et al. [32]. These investigators evaluated the effects of 400 mg of orally absorbed micronized progesterone or placebo in women at risk for premature labor; 88 % of the patients who were treated with oral micronized progesterone had decreased uterine activity compared with 42 % of the patients who received placebo. Furthermore Istwan et al. demonstrated that in patients with a history of one prior spontaneous preterm delivery prophylactic 17 OHP-C, confers a reduction in risk of preterm labour compared to untreated patients [33].

One of the main studies concerning the prophylactic use of vaginal progesterone to decrease the incidence of preterm birth in patients at a high risk, has been performed by Fonseca et al. [34]. Fonseca et al. demonstrated that the daily use of 100 mg vaginal progesterone is able to reduce the frequency of uterine contractions and the incidence of preterm birth. More recently it has also been demonstrated that the use of vaginal progesterone in asymptomatic women with a sonographic short cervix not only reduces the risk of preterm birth, but it is also able to reduce neonatal morbidity and mortality (Fig. 6.4, Table 6.2) [35]. In nulliparous women with a midtrimester cervical length <30 mm, the use of weekly intramuscular 17 OHP-C did not show a reduced incidence of preterm birth [36]. Comparing the use of vaginal

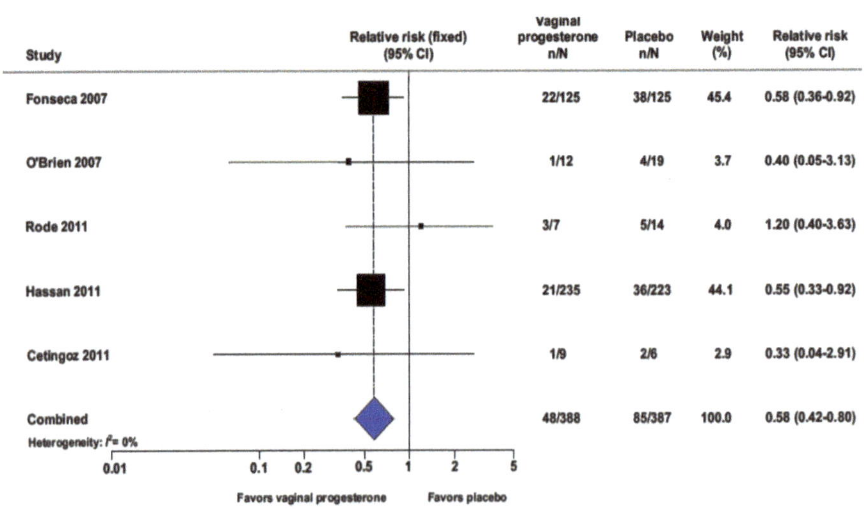

Fig. 6.4 Vaginal progesterone and reduction of preterm birth before 33 weeks of gestation: a metanalysis. From Romero R et al. Am J Obstet Gynecol 2012 [35]

Table 6.2 Vaginal progesterone in women with an aymptomatic short cervix in the midtrimester ultrasound decreases PTD and improves neonatal outcome

		No. of events/total no.				
Outcome	No. of trials	Vaginal progesterone	Placebo	Pooled RR (95% CI)	I^2 (%)	NNT (95% CI)
Preterm birth <37 wk	5	144/388	165/387	0.89 (0.75–1.06)	0	–
Preterm birth <36 wk	5	108/388	136/387	0.82 (0.67–1.00)	0	–
Preterm birth <35 wk	5	79/388	118/387	0.69 (0.55–0.88)	0	11 (7–27)
Preterm birth <34 wk	5	62/388	105/387	0.61 (0.47–0.81)	0	9 (7–19)
Preterm birth <30 wk	5	29/388	51/387	0.58 (0.38–0.89)	0	18 (12–69)
Preterm birth <28 wk	5	21/388	43/387	0.50 (0.30–0.81)	0	18 (13–47)
Spontaneous preterm birth <33 wk	5	39/388	71/387	0.57 (0.40–0.81)	0	13 (9–29)
Spontaneous preterm birth <34 wk	5	51/388	87/387	0.62 (0.46–0.84)	0	12 (8–28)
Respiratory distress syndrome	5	25/411	52/416	0.48 (0.30–0.76)	0	15 (11–33)
Neonatal death	5	8/411	15/416	0.55 (0.26–1.19)	43	–
Admission to NICU	5	85/411	121/416	0.75 (0.59–0.94)	0	14 (8–57)
Mechanical ventilation	5	35/411	51/416	0.66 (0.44–0.98)	0	24 (15–408)
Congenital anomaly	7	30/1967	34/1,954	0.89 (0.55–1.44)	0	–
Any maternal adverse event	3	86/624	80/595	1.04 (0.79–1.38)	0	–
Intraventricular hemorrhage	5	6/411	9/416	0.74 (0.27–2.05)	0	–

Modified from Romero et al. Am J Obstet Gynecol 2012 [35]

progesterone with the intramuscular route of administration, a statistically significant decrease of the number of admission to the Neonatal Intensive Care Unit and a significant deficiency of severe side effects has been demonstrated [37].

In both twin and triplet pregnancy, neither micronized progesterone nor 17 OHP-C have been shown to prevent preterm birth [29]. As long ago as 1980, Hartikainen-Sorri et al. reported a placebo controlled trial of 77 twin pregnancies treated during the last trimester until the 37th gestational week with weekly injections of either 17 OHP-C. There were no differences in the gestational length, birth weight or outcome of the neonates [38]. More recently, Combs et al. demonstrated that the prophylactic administration of 17 OH-C to mothers with twin pregnancy does not reduce the incidence of preterm delivery or neonatal morbidity [39].

4.1 Progesterone as a Possible Tocolytic Agent

The administration of high-dosage progesterone has been advocated as a possible tocolytic agent. Progesterone itself has been documented to inhibit uterine contractions, whereas 17 OHP-C seems to have no effect on uterine contractions (Fig. 6.1)

[29]. The action of progesterone is slow and it can be used for acute tocolysis only in conjunction with tocolytic agents such as β-agonists [33, 40]. The combination of the two drugs has shown synergistic effects by decreasing the need for high concentrations of β-agonists, which have potentially dangerous side effects (Table 6.3) [17, 22].

A recent study of Baumbach et al. demonstrates that progesterone significantly inhibits uterine contractility at relatively high concentrations and its combination with nifedipine and indomethacin increases this effect [41] (Fig. 6.5). Also Chanrachakul et al. recently demonstrated this synergetic effect of progesterone with β-mimetics. They concluded that progesterone is able to increase the relaxant effect of ritodrine, most likely through non-genic action [10].

Progesterone can be used for the maintenance of uterine quiescence in patients previously treated for an episode of threatened preterm labour. In comparison with nifedipine, progesterone has a significantly greater effect in prolonging pregnancies

Table 6.3 Use of progesterone in association with tocolytic agents

Patients	47	42
Gestational age	30.5 ws	30.3 ws
Treatment	– Ritodrine (100 mg in saline 0.1–0.3 mg/min)	– Ritodrine (50 mg in saline 0.1–0.3 mg/min) – Progesterone (200 mg die)
Outcome	– Deliveries after 48 h: 87 % – Deliveries After 7 days: 65 %	– Deliveries after 48 h: 85 % – Deliveries After 7 days: 68 %
Side effects	– Mat. tachicardia: 97 % – Nausea & vomiting: 28 % – Tremblings: 26 % – Palpitations: 32 % – Chest pain: 15 % – Hyperglycemia: 47 % – Hypokaliemia: 92 %	– Mat tachicardia: 42 % – Nausea & vomiting: 6 % – Tremblings: 12 % – Palpitations: 12 % – Chest pain: 8 % – Hyperglycemia: 28 % – Hypokaliemia: 23 %

Modified from Di Renzo et al. Curr Opin Obstet Gynecol 2005 [22]

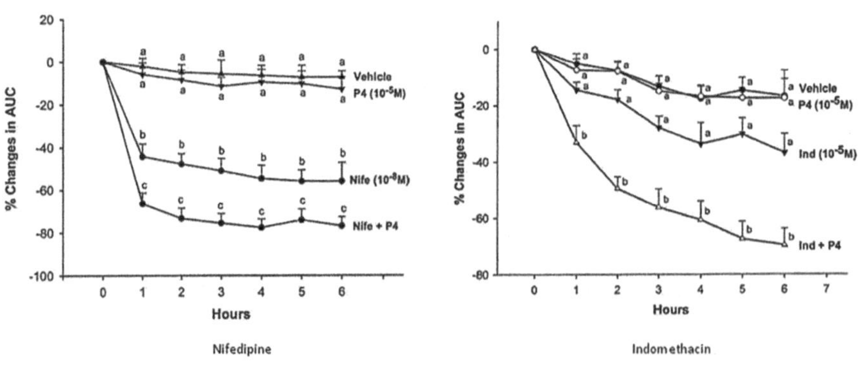

Fig. 6.5 Effects of nifedipine (*left*) or indomethacin (*right*) with and without P4 on myometrial contractility. *AUC* area under contractions curve, *Nife* nifedipina, *P4* progesterone. Modified from Baumbach J et al. Am J Obstet Gynecol 2012 [41]

of women with arrested preterm labour and, moreover, it results in a better neonatal outcome with less side effects [28]. Borna et al. have studied the role of progesterone in the maintenance of tocolysis after threatened preterm labour treated with intravenous magnesium sulphate. A longer latency period until delivery, a higher gestational age at delivery and lower frequency of low birth weight infants was found [42]. Moreover there is only evidence that micronized progesterone reduces the rate of preterm birth in women successfully treated for a previous episode of preterm labour [29].

Conclusions

In conclusion, progesterone is a molecule that has been thoroughly studied and evaluated. According to guidelines published in recent years by various societies (RCOG, ACOG, EAPM, SOGC) concerning the management of preterm labor, the use of progesterone can be recommended as follows: (1) early prophylaxis (from 12 to 36 weeks of gestation) with either micronized progesterone (100 mg vaginally daily) or 17 OHP-C (injection of 250 mg weekly) to prevent a recurrence in case of prior history of PTB [29]. (2) Vaginal micronized progesterone in singleton gestation, with and without history of prior PTB, and a short cervical length (<25 mm) at 18–23 weeks of gestation, since it has been found to be associated with reduction in the rate of PTB and perinatal morbidity and mortality (200 mg vaginal daily) [43, 44]. In this respect recently FIGO has published a best practice advice where it is recommended that women with a sonographic short cervix (< 25 mm) diagnosed in the mid-trimester should be offered daily vaginal micronized progesterone treatment for the prevention of preterm birth and neonatal morbidity. The progesterone formulation to be used is vaginal micronized progesterone (200 mg vaginal soft capsules) nightly or vaginal micronized progesterone gel (90 mg) each morning .Universal cervical length screening and vaginal progesterone treatment (90 mg vaginal gel or 200 mg micronized vaginal soft capsules) is a cost-effective model for the prevention of preterm birth [45]. (3) Vaginal micronised progesterone after one episode of threatened preterm labor (200 mg vaginal daily). (4) Vaginal progesterone in acute tocolysis (400 mg vaginal daily for 48 h) in conjunction with other tocolytic drugs usually used for tocolysis.

Paraphrasing Marcel Proust, "the real voyage of discovery consists not in seeking new landscapes, but in having new eyes," we can say that the rediscovery of progesterone activities place the hormone in a key role in the prevention of preterm birth.

References

1. Goldenberg RL, Culhane JF, Iams JD, Romero R. Epidemiology and causes of preterm birth. Lancet. 2008;371:75–84.
2. Haas DM. Preterm birth. Clin Evid. 2011, pii: 1404.

3. Di Renzo GC, Cutuli A, Liotta L, Burnelli L, Luzi G. Management of preterm labor: pharmacological and nonpharmacological aspects. In: Textbook of perinatal medicine. 2nd eds, Kurjak A, Chervenak FA. Informa Healthcare; 2006. Cap. 130, pp. 1394–1400.
4. Druckmann R, Druckmann MA. Progesterone and the immunology of pregnancy. J Steroid Biochem Mol Biol. 2005;97:389–96.
5. Szekeres-Bartho J, Barakonyi A, Par G, Polgar B, Palkovics T, Szereday L. Progesterone as an immunomodulatory molecule. Int Immunopharmacol. 2001;1:1037–48.
6. Di Renzo GC, Giardina I, Clerici G, Mattei A, Alajmi AH, Gerli S. The role of progesterone in maternal and fetal medicine. Gynecol Endocrinol. 2012;28:925–32.
7. Schwartz N, Xue X, Elovitz MA, Dowling O, Metz CN. Progesterone suppresses the fetal inflammatory response ex vivo. Am J Obstet Gynecol. 2009;201:211.
8. Fanchin R, Ayoubi JM, Olivennes F, Righini C, de Ziegler D, Frydman R. Hormonal influence on the uterine contractility during ovarian stimulation. Hum Reprod. 2000;15 Suppl 1:90–100.
9. Perusquía M, Jasso-Kamel J. Influence of 5alpha- and 5beta-reduced progestins on the contractility of isolated human myometrium at term. Life Sci. 2001;68:2933–44.
10. Chanrachakul B, Broughton Pipkin F, Warren AY, Arulkumaran S, Khan RN. Progesterone enhances the tocolytic effect of ritodrine in isolated pregnant human myometrium. Am J Obstet Gynecol. 2005;192:458–63.
11. Liu J, Matsuo H, Laoag-Fernandez JB, Xu Q, Maruo T. The effects of progesterone on apoptosis in the human trophoblast-derived HTR-8/SV neo cells. Mol Hum Reprod. 2007;13:869–74.
12. Czajkowski K, Sienko J, Mogilinski M, Bros M, Szczecina R, Czajkowska A. Uteroplacental circulation in early pregnancy complicated by threatened abortion supplemented with vaginal micronized progesterone or oral dydrogesterone. Fertil Steril. 2007;87:613–8.
13. Jacobsson B, Mattsby-Baltzer I, Andersch B, Bokström H, Holst RM, Wennerholm UB, Hagberg H. Microbial invasion and cytokine response in amniotic fluid in a Swedish population of women inpreterm labor. Acta Obstet Gynecol Scand. 2003;82:120–8.
14. Raghupathy R, Al Mutawa E, Makhseed M, Azizieh F, Szekeres-Bartho J. Modulation of cytokine production by dydrogesterone in lymphocytes from women with recurrent miscarriage. BJOG. 2005;112:1096–101.
15. Hudic I, Szekeres-Bartho J, Fatusic Z, et al. Dydrogesterone supplementation in women with threatened preterm delivery – the impact on cytokine profile, hormone profile, and progesterone-induced blocking factor. J Reprod Immunol. 2011;92:103–7.
16. Aisemberg J, Vercelli CA, Bariani MV, Billi SC, Wolfson ML, Franchi AM. Progesterone is essential for protecting against LPS-induced pregnancy loss. LIF as a potential mediator of the anti-inflammatory effect of progesterone. PLoS One. 2013;8:e56161. Epub 2013 Feb 7.
17. Di Renzo GC, Rosati A, Mattei A, Gojnic M, Gerli S. The changing role of progesterone in preterm labour. BJOG. 2005;112 Suppl 1:57–60.
18. Karteris E, Zervou S, Pang Y, Dong J, Hillhouse EW, Randeva HS, Thomas P. Progesterone signaling in human myometrium through two novel membrane G protein- coupled receptors: potential role in functional progesterone withdrawal at term. Mol Endocrinol. 2006;20: 1519–34.
19. Merlino AA, Welsh TN, Tan H, Yi LJ, Cannon V, Mercer BM, Mesiano S. Nuclear progesterone receptors in the human pregnancy myometrium: evidence that parturition involves functional progesterone withdrawal mediated by increased expression of progesterone receptor-A. J Clin Endocrinol Metab. 2007;92:1927–33.
20. Ruddok NK, Shi SQ, Jain S, Moore G, Hankins GDV, Romero R, Garfield RE. Progesterone, but not 17-alpha-hydroxyprogesterone caproate, inhibits human myometrial contractions. Am J Obstet Gynecol. 2008;199(391):e1–7.
21. Keirse MJ. Progestogen administration in pregnancy may prevent preterm delivery. Br J Obstet Gynaecol. 1990;97:149–54.
22. Di Renzo GC, Mattei A, Gojnic M, Gerli S. Progesterone and pregnancy. Curr Opin Obstet Gynecol. 2005;17:598–600.

23. Nahoul K, Dehennin L, Scholler R. Radioimmunoassay of plasma progesterone after oral administration of micronized progesterone. J Steroid Biochem. 1987;26:241–9.
24. Zava DT, Dollbaum CM, Blen M. Estrogen and progestin bioactivity of foods, herbs, and spices. Proc Soc Exp Biol Med. 1998;217:369–78.
25. Fitzpatrick LA, Good A. Micronized progesterone: clinical indications and comparison with current treatments. Fertil Steril. 1999;72:389–97.
26. Bulletti C, de Ziegler D, Flamigni C, Giacomucci E, Polli V, Bolelli G, Franceschetti F. Targeted drug delivery in gynaecology: the first uterine pass effect. Hum Reprod. 1997;12:1073–9.
27. Cicinelli E, de Ziegler D, Bulletti C, Matteo MG, Schonauer LM, Galantino P. Direct transport of progesterone from vagina to uterus. Obstet Gynecol. 2000;95:403–6.
28. Kamat S, Veena P, Rani R. Comparison of nifedipine and progesterone for maintenance tocolysis after arrested preterm labour. J Obstet Gynaecol. 2014;34(4):322–5. Epub ahead of print.
29. Di Renzo GC, Cabero Roura L, Facchinetti F, The European Association of Perinatal Medicine-Study Group on "Preterm Birth". Guidelines for the management of spontaneous preterm labour. Identification of spontaneous preterm labor, diagnosis of preterm premature rupture of membranes and preventive tools for preterm birth. J Matern Fetal Neonatal Med. 2011;24:659–67.
30. Goldstein P, Berrier J, Rosen S, Sacks HS, Chalmers TC. A meta-analysis of randomized control trials of progestational agents in pregnancy. Br J Obstet Gynaecol. 1989;96:265–74.
31. Daya S. Efficacy of progesterone support for pregnancy in women with recurrent miscarriage. A meta-analysis of controlled trials. Br J Obstet Gynaecol. 1989;96:275–80.
32. Erny R, Pigne A, Prouvost C, Gamerre M, Malet C, Serment H, Barrat J. The effects of oral administration of progesterone for premature labor. Am J Obstet Gynecol. 1986;154:525–9.
33. Barton JR, Barton LA, Istwan NB, Rhea DJ, Desch CN, Sibai BM. 17-Alphahydroxyprogesterone caproate in women with previous spontaneous preterm delivery: does a previous term delivery affect the rate of recurrence? Am J Obstet Gynecol. 2011;205:269.e1–6.
34. da Fonseca EB, Bittar RE, Carvalho MH, Zugaib M. Prophylactic administration of progesterone by vaginal suppository to reduce the incidence of spontaneous preterm birth in women at increased risk: a randomized placebo-controlled double-blind study. Am J Obstet Gynecol. 2003;188:419–24.
35. Romero R, Nicolaides K, Conde-Agudelo A, Tabor A, O'Brien JM, Cetingoz E, et al. Vaginal progesterone in women with an asymptomatic sonographic short cervix in the midtrimester decreases preterm delivery and neonatal morbidity: a systematic review and metaanalysis of individual patient data. Am J Obstet Gynecol. 2012;206:124.e1–19.
36. Grobman WA, Thom EA, Spong CY, Iams JD, Saade GR, Mercer BM, Eunice Kennedy Shriver National Institute of Child Health and Human Development Maternal-Fetal Medicine Units (MFMU) Network, et al. 17 Alpha-hydroxyprogesterone caproate to prevent prematurity in nulliparas with cervical length less than 30 mm. Am J Obstet Gynecol. 2012;207:390.e1–8.
37. Maher MA, Abdelaziz A, Ellaithy M, Bazeed MF. Prevention of preterm birth: a randomized trial of vaginal compared with intramuscularprogesterone. Acta Obstet Gynecol Scand. 2013;92:215–22.
38. Hartikainen-Sorri AL, Kauppila A. Tuimala R Inefficacy of 17 alpha-hydroxyprogesterone caproate in the prevention of prematurity in twin pregnancy. Obstet Gynecol. 1980;56:692–5.
39. Combs CA, Garite T, Maurel K, Das A, Porto M, Obstetrix Collaborative Research Network. 17-Hydroxyprogesterone caproate for twin pregnancy: a double-blind, randomized clinical trial. Am J Obstet Gynecol. 2011;204:221.e1–8.
40. Kurjak A, Chervenak FA. Management of preterm labor: pharmacological and non-pharmacological aspects. In: Textbook of perinatal medicine. 2nd ed. Informa Healthcare; 2006. Cap. 130, pp. 1394–1400.
41. Baumbach J, Shi SQ, Shi L, Balducci J, Coonrod DV, Garfield RE. Inhibition of uterine contractility with various tocolytics with and without progesterone: in vitro studies. Am J Obstet Gynecol. 2012;206:254.e1–5.
42. Borna S, Sahabi N. Progesterone for maintenance tocolytic therapy after threatened preterm labour: a randomised controlled trial. Aust N Z J Obstet Gynaecol. 2008;48:58–63.

43. Society for Maternal-Fetal Medicine Publications Committee, with assistance of Vincenzo Berghella. Progesterone and preterm birth prevention: translating clinical trials data into clinical practice. Am J Obstet Gynecol. 2012;206:376–86.
44. Farine D, Mundle WR, Dodd J, Basso M, Delisle MF, Farine D, Maternal Fetal Medicine Committee of the Society of Obstetricians and Gynaecologists of Canada, et al. The use of progesterone for prevention of preterm birth. J Obstet Gynaecol Can. 2008;30:67–77.
45. Miles RA, Paulson RJ, Lobo RA, Press MF, Dahmoush L, Sauer MV. Pharmacokinetics and endometrial tissue levels of progesterone after administration by intramuscular and vaginal routes: a comparative study. Fertil Steril. 1994;62:485–90.

Chapter 7
Abnormal Uterine Bleeding

Yochai Bar Shavit, Jaron Rabinovici, and Howard J.A. Carp

1 Introduction

Abnormal uterine bleeding is usually defined as vaginal bleeding in excess of 80 ml. Abnormal uterine bleeding (AUB) is one of the most frequent gynecological conditions accounting for 11–13 % of new gynecological consultations in reproductive age women and up to 70 % of new consultations in perimenopausal women [1]. AUB is usually seen in adolescents, and perimenopausally, but may also occur in the reproductive years. There is no clear consensus on the clinical evaluation of the patient with abnormal bleeding. Initial evaluation should assess the source and clinical features of the bleeding, and exclude organic causes bleeding such as fibroids, polyps, carcinoma of cervix or endometrium, coagulation defects, and systemic disease. In this chapter, the term AUB will be used rather than the previous terms of menorrhagia, dysfunctional bleeding etc. as the term AUB allows uterine bleeding abnormalities to be classified by bleeding pattern and etiology. Basically, AUB can result from three major reasons: Blood coagulation disorders, hormonal imbalances, and anatomical or malignant disorders. Based on these three rough divisions, the classification of AUB is known by the acronym PALM-COEIN. This acronym is based on the first letters of the major causes of AUB which can be either imaged or visualized (PALM: polyps, adenomyosis, leiomyoma and malignancy or hyperplasia). The others (which cannot be imaged) are known as COEIN (Coagulopathy

Y.B. Shavit, M.D.
Department of Obstetrics and Gynecology, Sheba Medical Center,
Tel Hashomer 52621, Israel

J. Rabinovici, M.D. • H.J.A. Carp, M.B.B.S., F.R.C.O.G. (✉)
Department of Obstetrics & Gynecology, Sheba Medical Center,
Tel Hashomer 52621, Israel

Sackler School of Medicine, Tel Aviv University, Tel Aviv, Israel
e-mail: carp@netvision.net.il

H.J.A. Carp (ed.), *Progestogens in Obstetrics and Gynecology*,
DOI 10.1007/978-3-319-14385-9_7

or AUB-C, ovulatory dysfunction or AUB-O, iatrogenic or AUB-I, endometrial factors AUB-E or not yet classified (AUB-N). There are three additional terms, to denote bleeding patterns, heavy menstrual bleeding (HMB), acute menstrual bleeding (AMB), and intermenstrual bleeding (IMB).

Progestogens are a significant factor in treatment. As outlined in other chapters, the choice of hormonal (progestogen) or surgical therapy varies according to the diagnosis, the patient's needs and fertility desire. This chapter will discuss the different types of progestogen in use for AUB as well as other treatment modalities.

2 Physiology of Menstruation

Lockwood [2] has given a full account of the pathophysiology of menstruation and the changes in AUB. Briefly, following menstruation, repair of the functional layer takes place. Stem cells in the endometrial stratum basalis start to produce a new functional layer under the influence of estrogen secreted by the ripening follicle. The endometrial epithelial and stromal cells proliferate. The stromal cells express vascular endothelial growth factor (VEGF) which induces angiogenesis [3] and the endothelial cells express angiopoietin-2 (Ang-2) [4]. In the secretory phase progesterone produced by the corpus luteum induces changes in the endometrium. Around the new blood vessels, progesterone augments expression of Angiopoietin-1 (Ang-1), from the stromal cells. Ang-1 stabilizes the vessels and blocks further angiogenesis. Progesterone also induces tissue factor (TF) mRNA and protein in the stromal cells [5]. TF is a receptor for coagulation factor VII and its active form, factor VIIa. TF initiates the clotting cascade, The cascade eventually leads to fibrin production. Decidualized stromal cells continue expressing TF throughout pregnancy [6] leading to the increased tendency to thrombosis in pregnancy. Progesterone also induces a second hemostatic protein, plasminogen activator inhibitor-1 (PAI-1). In addition to its anti-fibrinolytic properties PAI-1 restrains trophoblast invasion [7]. Hence, the luteal phase is associated with hemostatic, anti-fibrinolytic and antiproteolytic properties.

In the absence of pregnancy, luteal regression leads to progesterone withdrawal. The falling progesterone level leads to reduction of TF and PAI-1 expression [8]. When the falling level of progesterone reaches a threshold the spiral arteries in the endometrial stratum basalis tightly coil and constrict. The vasoconstriction leads to ischemia and necrosis in the functional layer.

Progesterone also inhibits expression of metaloproteinases 2, 3 and 9 (MMP-2, 3 and 9) expression. Progesterone withdrawal augments MMP expression by endometrial stromal cells. Progesterone withdrawal is also associated with up-regulation of the neutrophil and macrophage chemoattractants, interleukin-8 (IL-8) and macrophage chemoattractant protein-1, respectively [9]. Thus, progesterone withdrawal is associated with increased MMP expression and chemokines which promote leukocyte infiltration which add to the proteolytic milieu, promoting menstrual bleeding and tissue sloughing. As the spiral arteries relax, there is bleeding into the necrotic endometrium which together with the chemical changes in the endometrium lead to menstruation. After progesterone withdrawal, there is an increase in prostaglandin (PG) synthesis and a decrease in PG metabolism [10]. PG synthesis via COX-2 is particularly relevant in the vascular compartment, since this provides an explanation for the action of non-steroidal anti-inflammatory agents in the treatment of menstrual disorders including heavy and painful periods. Moreover, prostaglandin E (PGE) synergises with IL-8 to increase capillary permeability, which facilitates the efflux of leucocytes into the surrounding tissues [11].

2.1 *Pathophysiology in AUB*

AUB is due to derangements in the physiological processes described above. There are two principal causes of abnormal endometrial bleeding: impaired endometrial hemostasis and damage to endometrial vessels due to aberrant angiogenesis.

2.1.1 Pathophysiology of Anovulatory Bleeding

Anovulatory AUB is usually seen in adolescents and premenopausal women. In both cases there is bleeding from an endometrium which has been stimulated by estrogen, without progesterone modulation. In adolescents, the unopposed estrogen is often due to immaturity of the feedback mechanisms in the hypothalamic pituitary ovarian axis. There are various possibilities. If the negative feedback requires only a certain amount of estrogen to inhibit FSH secretion, falling FSH levels will lead to follicular degeneration and falling estrogen levels. In these circumstances, the endometrial shadow will be thin on ultrasound, and bleeding may be irregular in occurrence with polymenorrhea or acyclic bleeding. Consequently, the estrogen level never reaches high enough levels to release LH.

Another possibility is that the negative feedback requires higher than normal levels of estrogen to trigger the negative feedback. In these circumstances, the excess levels of unopposed estrogen may lead to hyperplasia and oligomenorrhea with prolonged heavy bleeding. Again, there is no positive feedback and LH release.

In the perimenopause, estrogen production is low compared to the reproductive years. Prolonged exposure to unopposed estrogen may also lead to endomerial hyperplasia, and prolonged heavy bleeding.

The mechanism of AUB in anovulation is due to estrogen breakthrough or withdrawal alone. There are none of the stabilising effects of constantly increasing estrogen levels or of post-ovulatory progesterone. While VEGF and Ang-2 are produced there is not enough Ang-1 to stabilise the vessels and block excess angiogenesis. There is no TF or PAI-1. Hence local blood clotting is sub optimal as is the anti fibrinolytic effect of PAI. When estrogen levels stay stable or fall, the endometrial lining cannot be maintained as it is estrogen dependent. In the absence of progesterone, there is no orderly constriction of the spiral arterioles, and no orderly necrosis of the functional endometrium. The bleeding therefore occurs from excess of fragile blood vessels, with suboptimal thrombosis to stop the bleeding, and possibly excessive fibrinolysis.

2.1.2 Bleeding Dyscrasias

Bleeding diatheses generally present as heavy menstrual bleeding commencing at menarche and are present in 10.7 % of patients with HMB compared to 3.2 % of control women [12]. Von Willebrand's disease is the most prevalent defect associated with HMB with a prevalence of 5–20 % [13]. Screening includes activated partial thromboplastin time (aPTT) and ristocetin cofactor assay. Treatment consists of combined hormonal contraceptives which presumably induce TF and PAI-1 levels to compensate for the hemostatic defect.

3 Clinical Features

AUB usually occurs at the extremes of reproductive life, in adolescents and at the peri-menopause. However, AUB can also occur in the reproductive years. At the extremes of reproductive life, AUB is mainly associated with anovulation. At puberty, the negative feedback matures before the positive feedback. Therefore, the first few cycles have FSH secretion and subsequent estrogen production from ripening follicles. However, the negative feedback inhibits further FSH secretion, so the follicles cannot produce sufficient estrogen to trigger the positive feedback and LH production. The bleeding from an endometrium which has been stimulated by estrogen without endogenous or exogenous progesterone modification is due to a fragile, vascular endometrium, that lacks sufficient stromal support. In estrogen withdrawal bleeding, as one area of bleeding begins to heal, another area begins to slough, which results in erratic bleeding patterns and may manifest as prolonged and heavy and often irregular bleeding (HMB, IMB). In the perimenopause there is a diminished estrogen response to FSH. Again insufficient estrogen is produced, so that the positive feedback is not triggered. In all forms of AUB, the role of clinical examination and imaging are paramount to exclude organic disease. In the adolescent, the likelihood of organic disease such as malignancy is low. Clinical abdominal examination, usually gives little information as to cause. Additionally, if the adolescent is a virgin, vaginal examination is inappropriate. Consequently, if imaging is normal, there is probably little need to rule out organic disease. However, in the perimenopausal patient, the chance of organic disease is high, and there is a need for clinical examination to assess uterine size, speculum examination of the cervix, and cervical cytology to exclude malignant changes. Additionally, clinical examination is insufficient, and more invasive diagnostic techniques such as endometrial biopsy may be indicated.

The patient presents for consultation, because the amount of bleeding seems abnormal for her. There may not necessarily be more than 80 ml of blood lost, but it is necessary to accept the patients subjective distress of an abnormally perceived bleeding pattern. A quantitative estimate of the amount of bleeding can be obtained by a pictorial blood loss assessment chart. In this chart, the patient uses a points system for the amount that pads or tampons are soaked in blood, and for the number of days of bleeding. However, charting is often difficult to apply in clinical practice. Recently, smartphone applications have become available, in which the patient can record the amount of bleeding.

4 Diagnosis

The main step in the solution of clinical problems is the correct diagnosis of the disorder. A correct diagnosis of AUB begins with a clinical gynaecological evaluation of the integrity of the pelvic organs. Following this step further imaging and laboratory examination are in order to arrive at the right diagnosis and the right choice of therapy.

4.1 Importance of Imaging

Ultrasound, MRI, Hysteroscopy, and sonohysterography are used to image the uterus. These techniques are invaluable for making a diagnosis and directing treatment. Hysterosalpingography (HSG) was used in previous years. Although HSG can detect intra-cavity lesions such as fibroids or polyps, it is poor at differentiating between atrophy or hyperplasia.

4.2 Ultrasound

The role of ultrasound is to detect pathology not detectable at a clinical examination (e.g. endometrial pathology, such as endometial thickness, (hypoplasia, or hyperplasia), small submucous myomas, adenomyosis, polyp, etc. Figure 7.1 shows a sonogram of an atrophic or hypoplastic endometrium. Figure 7.2 shows a sonogram of endometrial hyperplasia. The ultrasound examination is also used to confirm or refute a diagnosis suspected on the basis of abnormal findings at palpation (e.g. uterine intramural or subserous myomas, or adnexal masses). The clinician then needs to decide if an abnormal ultrasound finding is the likely cause of the abnormal bleeding or if it is an incidental finding unrelated to the woman's symptoms. An ultrasound examination carried out because of abnormal vaginal bleeding should also include examination of the adnexa and the urinary bladder, as abnormal bleeding may be explained by a hormone-producing ovarian tumour or a tumour in the urinary bladder (i.e. if the woman confuses bleeding from the urethra with vaginal bleeding). US can also diagnose rare A-V malformations.

If a polyp, adenomyosis or leiomyoma are found, the treatment is surgical or interventional (uterine artery embolization for unterine fibroids or magnetic resonance guided highly focused ultrasound for uterine fibroids and adenomyosis)

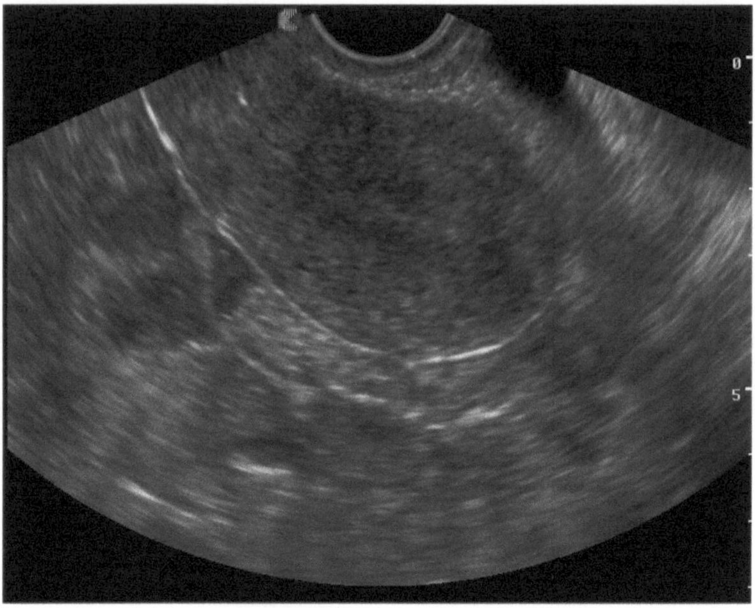

Fig. 7.1 Endometrial atrophy. Endometrial atrophy. The endometrial shadow can be seen as a thin feint line

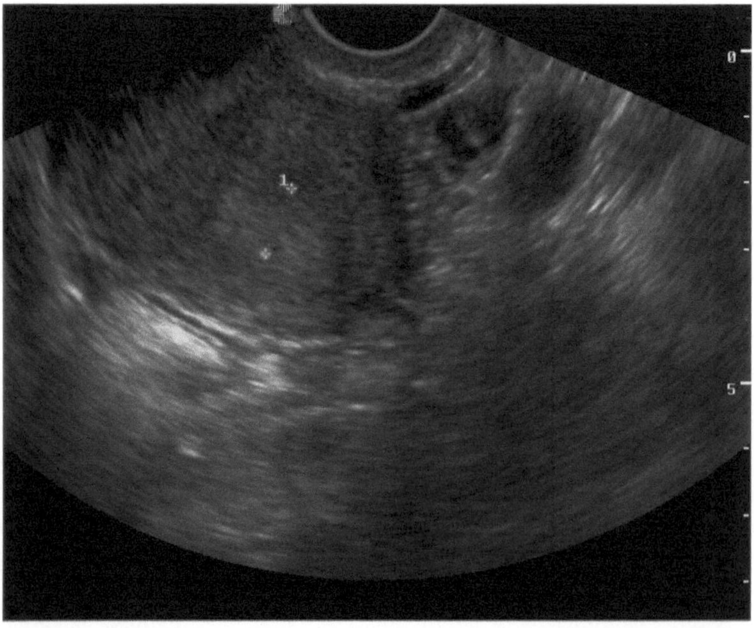

Fig. 7.2 Endometrial hyperplasia. Endometrial hyperplasia. The thickened endometrial shadow can be seen between the two calipers

and therefore outside of the scope of this chapter. If no abnormalities are found, endometrial biopsy should be considered. The big advantage of ultrasound, is that it can be performed on adolescents with an abdominal probe, thus precluding the necessity for vaginal or speculum examination. Doppler ultrasonography may provide additional information for characterizing endometrial and myometrial abnormalities, particularly arterio-venous malformations.

4.3 Hydrosonography

This technique is also known as saline-contrast sonohysterography, saline infusion sonography (SIS) or sonohysterography. In this technique saline is infused into the uterine cavity during transvaginal scanning. SIS separates the uterine walls and thereby clarifies the presence of focal lesions protruding into the uterine cavity [14]. As long as the cervical canal is not stenotic, SIS is relatively easy to perform. After inserting the catheter into the uterine cavity and the ultrasound probe into the vagina, a few millilitres of saline are injected. Virtually all endometrial pathology grows focally in the uterine cavity [15]. If no focal lesions are present in the uterine cavity, the odds of malignancy decrease 20-fold, and the odds of any endometrial pathology decrease 30-fold [15]. A smooth endometrium at SIS is a strong sign of normality.

As most focal lesions cannot be removed at all, or only be partially removed by blind endometrial sampling, such as pipelle, or dilatation and curettage, focal lesions should be hysteroscopically resected under direct visual control.

4.4 Magnetic Resonance Imaging (MRI)

MRI is not generally recommended as a first-line procedure for investigating AUB. MRI is a good second line procedure if ultrasound reveals a bulky, polymyomatous uterus, or if adenomyosis is suspected. MRI has the advantage of distinguishing between myomas, sarcomas, and adenomyosis. Therefore, MRI can also optimize treatment strategy regarding the use of major surgery, or minimally invasive procedures. MRI can also provide a diagnostic assessment of the endometrium when the uterine cavity is inaccessible [16].

5 Hysteroscopy

Hysteroscopy has a few unique advantages over other forms of imaging, despite the need for a skilled endoscopic surgeon. Diagnostic hysteroscopy can be performed as an office procedure and permits direct visualization of the endometrial cavity. Therefore, it can be very valuable for diagnosing endometrial focal lesions, atrophy and hyperplasia as well as primary or secondary malformations of the uterine cavity (septum, polyp, residua, caesarean section niche, etc). Hysteroscopy also has the advantage of allowing a targeted biopsy to be taken, particularly in focal lesions which may be missed by blind endometrial sampling techniques. The likelihood of endometrial cancer diagnosis after a negative hysteroscopy result is 0.4–0.5 % [17]. In some settings, especially if uterine pathology is suspected, a diagnostic hysteroscopy can be combined with an operative procedure in the same session if required. Thc European guidelines [18] suggest that hysteroscopy is a second line procedure when ultrasound suggests a focal lesion, when biopsy is non diagnostic, or as an operative procedure if medical treatment fails after 3–6 months.

6 Biopsy

Histological examination is usually thought to be the gold standard for making a diagnosis of uterine pathology. Endometrial sampling for the diagnosis or exclusion of mostly hormonally induced endometrial changes (hyperplasia or endometrial cancer) is most often performed with a pipelle. The biopsy also may provide information about the hormonal status of the endometrium. However, there is a 0–54 % rate of sampling failure [19]. Another important limitation of pipelle biopsy is that it samples an average of only 4 % of the endometrium with a reported range of 0–12 % [20]. Usually a polyp is an incidental finding during endometrial sampling and most often is not entirely removed by pipelle.

Classically, endometrial sampling was performed by dilatation and curettage (D&C). Pipelle biopsy has replaced D&C, as pipelle biopsy is an office procedure, thus less invasive and less expensive than D&C. In addition, pipelle biopsy is safer, as the general anesthetic usually used in D&C is not required. D&C has been reported to lack the ability to identify uterine focal lesions [21], and blind excision of focal lesions by curettage may be incomplete.

The question arises as to when sampling is indicated. In the adolescent, there is little place for biopsy, unless absolutely necessary and might be considered only in extreme cicumstances.

Goldstein [22] summarised five large prospective studies in women with postmenopausal women. An endometrial thickness of ≤4 mm on transvaginal ultrasound with bleeding was associated with a risk of malignancy of 1 in 917 (3 cancers in 2,752 patients). Goldstein [22] concluded that in postmenopausal bleeding, biopsy is not indicated when the endometrial thickness is ≤4 mm. Furthermore, if biopsy is performed in patients with a thin endometrium, it is most likely that no tissue would be obtained for histology. In a study of 97 consecutive patients with post menopausal bleeding evaluated by endometrial biopsy, only 82 % of the patients with an endometrial thickness ≤5 mm (n = 45) had a successful Pipelle biopsy completed [23], and only 27 % of them produced a sample which was adequate for diagnosis. The results on postmenopausal women can be extrapolated to premenopausal women. However, in women with endometrial hyperplasia, endometrial sampling is indicated [24, 25], as there is a high possibility of malignancy.

Endometrial sonographic thickness as an indicator of the need for biopsy is problematic in premenopausal women with AUB, as endometrial thickness changes throughout the menstrual cycle (from 4 to 8 mm during the proliferative phase, to 8–14 mm during the secretory phase [26]). In our hands, the use of hormonal assessment prior to endometrial sampling has proven very to be clinically important: Determination of blood estradiol, progesterone and beta-hCG levels prior to endometrial sampling can avoid sampling of a pregnant or postovulatory endometrium and allow re-assessment of the ultrasound findings and endometrial thickness in view of the hormonal state of the patient.

7 Principles of Treatment

Treatment has a number of objectives, to lessen or stop the bleeding, and to provide long term relief. AUB due to a structural problem (polyp, adenomyosis, and leiomyoma, etc.) are treated surgically, and therefore outside the scope of this chapter. Medical management of endometrial cancer has been described in Chap. 12. The primary goal of medical therapy should be to stabilize the endometrium with estrogen to provide initial hemostasis, followed by combined estrogen/progestogens for endometrial stability and induction of a menstruation-like withdrawal bleed. This induced bleeding can at first be strong because of a medical "curettage" of a thickened endometrial layer and is usually limited in time, especially if hormonal therapy is continued afterwards. However, this basic plan of action should be modified according to the patients needs, desire for fertility, anemia, endometrial thickness etc. In addition to hormonal therapy, other medical treatment modalities are available such as NSAIDS, tranexamic acid, and receptor modulators etc. Some specific modifications are listed below.

7.1 Adolescent AUB

The presence of irregular and heavy AUB is a relatively common cause of concern among adolescents and their parents, as well as a frequent cause of visits to emergency departments, gynaecologists, and pediatricians [1]. Adolescent patients pose particular diagnostic problems because the characteristics of normal puberty often overlap with signs and symptoms of PCOS. The aims of AUB treatment in adolescents is therefore to stop bleeding, prevent or reverse anemia and achieve adequate cycle control. Anemia can be corrected with iron supplementation, with added folate and vitamin B12 if necessary, but these alone will not address the underlying cause of anemia: bleeding. Bleeding can usually be controlled with combined oral contraceptive pills (OCPs) taken continuously for several months. Alternatively, monophasic OCPs containing 20–30 μg of ethinyl estradiol and a relatively androgenic progestogen such as 0.3 mg of norgestrel or 0.15 mg of levonorgestrel can be used cyclically. If breakthrough bleeding occurs, or heavy menstrual bleeding persists and other causes of

AUB have been excluded, the dose can be doubled for a short period of time to two pills per day. Since combined hormonal contraceptives can increase levels of coagulation factors such as factor VIII and von Willebrand factor, OCP's might have an additional effect in cases of an underlying coagulopathies.

If estrogen is contraindicated due to a history of thrombosis, migraine, hypertension etc., progestogens alone can be used. Examples are: oral medroxyprogesterone acetate (MPA), or norethisterone (norethindrone acetate, NETA). Oral MPA 10 mg daily or NETA 5 mg daily given for 10–14 days each month generates a secretory endometrium that induces a withdrawal bleed 1–7 days after stopping the medication. However, (NETA) can be aromatised to ethinylestradiol [27]. Kuhnz et al. [28] reported that this conversion resulted in a dose that was equivalent to taking 4–6 μg of ethinyl estradiol for each 1 mg of NETA ingested. The conversion ratio of NETA to EE has been subsequently estimated to be between 0.2 and 0.33 % for different doses [29], Chu et al. [30] concluded that a daily dose of 10–20 mg NETA equates to taking a 20–30 μg ethinyl estradiol COC, Conversion to estrogen and the estrogenic effects are of no relevance when these progestogens are taken in low-dose progestogen-only, or combined oral contraceptive pills [31] but probably explains why high-dose norethisterone and its ester are effective at delaying and regulating menstrual bleeding. There are no similar implications for other progestogens in either low or high doses, since these structural issues do not apply [32–34].

It has been reported in other chapters in this book that dydrogesterone binds the progesterone receptor up to 50 % more than progesterone itself. However, dydrogesterone stimulates the progesterone receptor alone. It may therefore be appropriate in patients with a thickened endometrium in whom progesterone only effects are required. However, if there is a thin endometrium, estrogen will also be required to provide hemostasis, in addition to dydrogesterone.

The LNG-IUS or etonogestrel/ethinyl estradiol vaginal ring are other possibilities, but are not usually acceptable in adolescents. Clomiphene citrate has occasionally been used in anovulatory adolescents. Clomiphene is a SERM which blocks the estrogen receptor in the hypothalamus, thus inhibiting the negative feedback. Hence there is up-regulation of the hypothalamic-pituitary-gonadal axis. Subsequently, estrogen levels can rise to the level required to induce LH release. The use of clomiphene has recently been reported again as a possible therapy in anovulatory adolescents [35]. During the use of clomiphene citrate in adolescents the chance of conception and the rare possibilities of side-effects (headaches, vision changes, ovarian hyperstimulation, etc.) should be taken into consideration. Further, studies demonstrated that prolonged use of clomiphene might be associated with ovarian cancer.

7.1.1 Acute Uterine Bleeding

In the case of acute AUB, there is a need to stop the bleeding immediately. Hormonal management is the first line of medical therapy if there are no known or suspected bleeding disorders. Treatment options include oral estrogenic compounds, IV conjugated equine estrogen, combined oralcontraceptives (OCs), and oral progestins.

If AUB can result from estrogen breakthrough bleeding then a logical approach is to stop the process by adding estrogens orally or parenterally. This approach usually leads to a very quick cessation of bleeding and after several days without bleeding, combined estrogen and progestagen treatment can be administered until the planned time of the next menstruation.

Intravenous conjugated equine estrogen is approved by the US Food and Drug Administration (FDA) for the treatment of acute AUB. A dose of 25 mg can be given intravenously. However, the patient requires cycling by progestogen immediately afterwards. In one randomized controlled trial of 34 women, IV conjugated equine estrogen stopped bleeding in 72 % of participants within 8 h of administration compared with 38 % of participants treated with a placebo [36]. Little data exist regarding the use of IV estrogen in patients with cardiovascular or thromboembolic risk factors.

High doses of oral contraceptive pills as well as MPA have been used with variable success [37]. For COC large doses such as three pills per day may be required with the resulting side-effects of large doses of hormones. However, estrogen only is probably more effective than COC. Progestogens can be used as sole agents in acute AUB. MPA can be given up to 20 mg three times daily for a week or NETA can be given in doses up to 40 mg daily in divided doses until bleeding stops and then the dose can be tapered down [38]. Another treatment regimen for acute AUB is depo-medroxyprogesterone acetate 150 mg given intramuscularly followed by MPA 20 mg given orally thrice daily for 3 days. This treatment stopped bleeding within 5 days in all 48 women in a pilot study. Study participants reported infrequent side effects and high satisfaction.

Monro et al. [37] compared the results of oral contraceptives three times daily for 1 week with MPA administered three times daily for 1 week. The study found that bleeding stopped in 88 % of women who took OCs and 76 % of women who took MPA within a median time of 3 days.

The author uses a regimen of oral synthetic estradiol (Estrofem) 2–4 mg followed by oral MPA 10 mg or micronized progesterone 200 mg nocte. This regimen results in good relief of acute AUB.

7.2 *Perimenopausal Bleeding*

Perimenopausal patients need to have organic disease excluded before medical therapy is prescribed. As mentioned above, there is a higher incidence of organic disease in this age group. As in the other age groups, anemia may need to be corrected. Perimenopausal women with AUB, may be treated with cyclic progestin therapy, low-dose oral contraceptive pills, the levonorgestrel IUD, or cyclic hormone therapy. Each treatment modality has advantages and disadvantages. The OCP and LNG-IUS provide contraception, in addition to hemostasis. Estrogen therapy, also provides relief from perimenopausal symptoms, such as hot flashes, night sweats, and vaginal atrophy. The choice of therapy often is guided by the patient's priorities.

Endometrial thickness will also give an idea of whether estrogen is required, or whether the patient can be managed on progestogen alone. In a study of 120 perimenopausal women treated by continuous estrogen and cyclic progestin or cyclic progestogen alone [39], 86 % of women in the combined treatment group experienced cyclic menstrual bleeding, and reduced vasomotor symptoms. In addition, 76 % of the women rated their bleeding as normal in amount and duration.

7.2.1 LNG-IUS

The LNG-IUS is particularly useful for the treatment of heavy menstrual bleeding in women who desire contraception. The LNG-IUS has been shown to be the most effective treatment in reducing menstrual blood loss compared with other medical therapies for chronic AUB and can reduce menstrual blood loss by more than 80 % and even induce hormonal amenorrhea [40]. However, the LNG-IUS takes time to achieve adequate endometrial quiescence. Initially there may be increased bleeding. The LNG-IUS results in greater increases in hemoglobin and serum ferritin levels after 6 months compared with oral MPA from day 16 to 26 of the menstrual cycle [41]. Women also reported higher rates of subjective improvement in their bleeding despite the known initial side effect of irregular bleeding after LNG-IUS insertion. The efficacy of the levonorgestrel IUS was evaluated by Vilos et al. [42] in 56 obese perimenopausal women with AUB. The mean age was 42 years and the mean body mass index was greater than 30. At the 48-month follow-up, the satisfaction rate was 75 %; amenorrhea and hypomenorrhea were noted with longer use. Hence, The LNG-IUS is an excellent long-term treatment modality for heavy menstrual bleeding when contraception is also required.

7.2.2 Endometrial Hyperplasia

Endometrial hyperplasia, whether simple or complex, with or without atypia has malignant potential. Figure 7.2 shows a sonogram of endometrial hyperplasia. If atypia is present, there is a 29 % risk of progression to endometrial cancer [43]. However, in simple hyperplasia, the risk can be as low as 1 %. Therefore treatment should aim at eradicating the condition. In the perimenopausal or post menopausal woman, hysterectomy is probably the best treatment option. However, in younger women, endometrial hyperplasia can be found in anovulatory cycles, polycystic ovary syndrome, or obesity. If fertility is desired, progestogens are the mainstay of treatment. In hyperplasia, there is little need for estrogen as the condition is due to excess stimulation with unopposed estrogen. Therefore, progestogens can be used to convert the endometrium to a secretory pattern. Once endometrial hyperplasia has been diagnosed, it is essential to repeat the biopsy 3–6 months afterwards to confirm that regression has taken place. However, it has been reported that the median length of progestin treatment required for regression can be 9 months. Additionally endometrial hyperplasia is closely related to insulin resistance and metabolic disorder. A low body mass index of <35 kg/m^2 has been reported to be associated with a high resolution rate in patients receiving progestogens [44]. There are case reports which show that if there is no response to progestogens, reversal of hyperplasia could be induced with metformin in addition to progestogens [45, 46].

The main progestogens for treating endometrial hyperplasia are megestrol acetate, medroxyprogesterone 17-acetate [MPA], dydrogesterone and the LNG-IUS. However, there is no consensus on dose, treatment, duration, route of administration, or the most effective progestogen [47]. No evidence is available from randomized controlled trials. Hence treatment is somewhat empiric, and administered in a trial and error fashion. The overall response rate has been reported to be approximately 70 %. Moreover, oral progestins are associated with poor compliance and systemic side effects that may limit overall efficacy [48]. Below are some of the advantages and disadvantages of each regimen.

Megestrol has antiestrogenic and anti androgenic effects, which may not be acceptable to the patient, or may decrease compliance. Megestrol acetate also has glucocorticoid effects. Symptoms of Cushing's syndrome, steroid diabetes, and adrenal insufficiency, have been reported with the use of megestrol acetate (in the medical literature, albeit sporadically) [49].

Medroxy progesterone acetate (MPA) however, is an agonist of the progesterone, androgen, and glucocorticoid receptors [50]. Hence there may be side effects of acne and hirsutism in some patients. MPA has glucocorticoid properties, and as a result can cause Cushing's syndrome, steroid diabetes, and adrenal insufficiency. Mesci-Haftac et al. [51] assessed 69 patients with simple hyperplasia, who received MPA. Hyperplasia persisted in 19.7 %. The regression rate to benign findings was 77.1 %. Atypia and progression to complex hyperplsia occurred in 3.2 % of the patients.

Dydrogesterone has been available for over 40 years, consequently, the reports of this progestin go back a long time. In 1988, Meden-Vrtovec and Hren-Bozic [52] reported on 50 patients with cystic glandular hyperplasia who were treated with dydrogesterone for six cycles in a dose of 20–30 mg. Repeat curettage showed persistence of the hyperplasia in 5 of 18 patients (28 %).

Micronized progesterone was introduced in order to provide a bio-identical form of progesterone. Due to metabolism in the liver, it is often administered vaginally, allowing it to bypass the liver, and have an increased local concentration in the endometrium. Tasci et al. [53] compared sixty premenopausal women with endometrial hyperplasia without atypia in a prospective controlled study. Group I included 30 patients who received lynestrenol in a dose of 15 mg per day while Group II included 30 patients who received micronised progesterone 200 mg per day for 12 days per cycle for 3 months. After 3 months of treatment no patient in either group showed progression of the hyperplasia. In the lynestrenol group, the rate of resolution was higher than in the micronized progesterone group ($p=0.045$). Lynestrenol was more effective in inducing resolution in patients more than 45 years ($p=0.036$). The authors concluded that lynestrenol ensures better endometrial control than micronized progesterone at the above doses in simple hyperplasia without atypia.

The Levonorgestrel Intrauterine device has also been used to treat endometrial hyperplasia. The LNG-IUS has been compared to other progestogens: Dydrogesterone, NETA and MPA. El Behery et al. [54], assessed the results of 138 women aged between 30 and 50 years with AUB and hyperplasia who were randomized to receive either LNG-IUS or dydrogesterone for 6 months. The outcome measures were regression of hyperplasia, and side effects or recurrence during the follow-up period. After 6 months of treatment, regression of hyperplasia occurred in 96 % of women in the LNG-IUS group versus 80 % of women in the dydrogesterone group ($P<0.001$). Intermenstrual spotting and amenorrhea were more common in the LNG-IUS group ($p=0.01$ and 0.0001, respectively). Patient satisfaction was significantly higher in the LNG-IUS group ($P=0.0001$). Hysterectomy rates were lower in the LNG-IUS group than in the dydrogesterone group ($p=0.001$). The recurrence rate was 0 % in the LNG-IUD group compared to 12.5 % in the dydrogesterone group. Gallos et al. [55] reported similar results.

The LNG-IUS has also been compared to NETA [56]. One hundred and twenty-nine perimenopausal women with non-atypical endometrial hyperplasia were assessed in a randomized controlled trial. Patients received either the LNG-IUS or NETA for 3 weeks per cycle for 3–6 months. A significantly higher regression rate was noted in the LNG-IUS group than in the NETA group (79.7 % vs. 60.7 %, RR, 1.31 after 6 months). However, no significant difference was found regarding the median time to regression (3 months). The hysterectomy rate during the follow-up period was significantly higher in the NETA group (57.4 % vs.22 %, $P<0.001$).

In a prospective RCT [57] comprising 90 premenopausal women with a histological diagnosis of simple endometrial hyperplasia without atypia, patients were randomly allocated to three groups of 30 patients. One group received MPA, 10 mg per day. The second group received NETA, 15 mg per day for 10 days per cycle. The third group had a LNG-IUS inserted. Patients were reevaluated after 3 months of treatment. Patients with regression and persistence were offered the same medication they were using for another 3 months. Patients in the LNG-IUS group showed the highest resolution rate (66.67 %). Patients in the MPA and NETA groups had a resolution rate of 36.66 and 40 %, respectively. The patients with a LNG-IUS showed a regression rate of 33.3 %, whereas patients receiving MPA and NET showed a regression (and persistence) rate of 60 and 56.67 %, respectively There was a statistically significant difference between the three groups regarding the proportion of patients requiring further treatment for another 3 months ($\chi^2=6.501$; $P=.0387$) in favour of the LNG-IUS.

7.3 Receptor Modulators

Receptor modulators have been used in preliminary trials and have shown promising results. The selective estrogen receptor modulator (SERM), ormeloxifene has an anti-estrogenic effect, which retards endometrial maturation. In a randomized study, Ravibabu et al. [58], reported that ormeloxifene decreases blood loss by 90 % and that the reduction was statistically significant ($P<0.001$). There was also significant decrease in the mean endometrial thickness ($P < 0.001$) after treatment with ormeloxifene when compared to mean baseline value. There was significant improvement, 84 % of patients had relief from dysmenorrhoea ($P < 0.001$), but anti-estrogenic side effects such as hot flashes and vaginal dryness are a cause for concern.

Mifepristone (Ru486), is a selective progesterone receptor modulator (SPRM). Mifepristone induces amenorrhea whilst maintaining endogenous estrogen secretion. Amenorrhea is caused by complete binding of the progesterone receptor, causing atrophy of spiral arteries and hence, anovulatory amenorrhea [59].

7.4 NSAIDS

A certain amount of relief may be obtained from non steroidal anti-inflammatory drugs (NSAIDs). NSAIDs inhibit the enzyme cyclooxygenase, thereby reducing the raised PG levels which are found in women with heavy menstrual bleeding. A recent Cochrane review [60] showed that NSAIDs (mefenamic acid, naproxen, ibuprofen, flurbiprofen, meclofenamic acid, diclofenlac, indomethacin, and acetylsalicylic acid) are more effective than placebo in reducing menstrual blood loss in women with regular menstrual cycles, but they less effective than tranexamic acid, danazol, or the levonorgestrel-releasing intrauterine device (LNG-IUS).

7.5 Antifibrinolytic Drugs

Tranexamic acid is a lysine analogue that allows the formation of stable blood clots by preventing fibrin filament breakdown without influencing coagulation in healthy blood vessels (NICE). Data from controlled clinical trials indicate reduced bleeding in AUB by 30–55 % [61, 62]. However, there are side effects including headache, nausea and vomiting. Tranexamic acid is the only nonhormonal, noncontraceptive agent approved by the U.S. Food and Drug Administration (FDA) for the treatment of HMB. Concerns have been raised about the potential for thromboses during tranexamic acid treatment. However, population-based data do not indicate an increased risk of thromboses [63]. Tranexamic acid is most often used for women in their reproductive years to reduce heavy menstrual bleeding. One potential benefit is that tranexamic acid is only used during menstruation rather than continuously or for the majority of the menstrual cycle.

References

1. Spencer CP, Whitehead MI. Endometrial assessment re-visited. Br J Obstet Gynaecol. 1999;106:623–32.
2. Lockwood CJ. Mechanisms of normal and abnormal endometrial bleeding. Menopause. 2011;18(4):408–11.
3. Krikun G, Schatz F, Lockwood CJ. Endometrial angiogenesis: from physiology to pathology. Ann N Y Acad Sci. 2004;1034:27–35.

4. Schatz F, Krikun G, Caze R, Rahman M, Lockwood CJ. Progestin-regulated expression of tissue factor in decidual cells: implications in endometrial hemostasis, menstruation and angiogenesis. Steroids. 2003;68:849–60.
5. Lockwood CJ, Nemerson Y, Guller S, Krikun G, Alvarez M, Hausknecht V, Gurpide E, Schatz F. Progestational regulation of human endometrial stromal cell tissue factor expression during decidualization. J Clin Endocrinol Metab. 1993;76:231–6.
6. Lockwood CJ, Murk W, Kayisli UA, Buchwalder LF, Huang ST, Funai EF, Krikun G, Schatz F. Progestin and thrombin regulate tissue factor expression in human term decidual cells. J Clin Endocrinol Metab. 2009;94:2164–70.
7. Zini JM, Murray SC, Graham CH, Lala PK, Karikó K, Barnathan ES, Mazar A, Henkin J, Cines DB, McCrae KR. Characterization of urokinase receptor expression by human placental trophoblasts. Blood. 1992;79:2917–29.
8. Papp C, Schatz F, Krikun G, Hausknecht V, Lockwood CJ. Biological mechanisms underlying the clinical effects of mifepristone (RU 486) on the endometrium. Early Pregnancy. 2000;4:230–9.
9. Critchley HO, Jones RL, Lea RG, Drudy TA, Kelly RW, Williams AR, Baird DT. Role of inflammatory mediators in human endometrium during progesterone withdrawal and early pregnancy. J Clin Endocrinol Metab. 1999;84:240–8.
10. Baird DT, Cameron ST, Critchley HO, et al. Prostaglandins and menstruation. Eur J Obstet Gynecol Reprod Biol. 1996;70(1):15–7.
11. Critchley HOD, Kelly RW, Brenner RM, et al. The endocrinology of menstruation – a role for the immune system. Clin Endocrinol. 2001;55:701–10.
12. Dilley A, Drews C, Miller C, Lally C, Austin H, Ramaswamy D, Lurye D, Evatt B. von Willebrand disease and other inherited bleeding disorders in women with diagnosed menorrhagia. Obstet Gynecol. 2001;97:630–6.
13. James A, Matchar DB, Myers ER. Testing for von Willebrand disease in women with menorrhagia: a systematic review. Obstet Gynecol. 2004;104:381–8.
14. Parsons AK, Lense JJ. Sonohysterography for endometrial abnormalities: preliminary results. J Clin Ultrasound. 1993;21:87–95.
15. Epstein E, Ramirez A, Skoog L, et al. Transvaginal sonography, saline contrast sonohysterography and hysteroscopy for the investigation of women with postmenopausal bleeding and endometrium >5 mm. Ultrasound Obstet Gynecol. 2001;18:157–62.
16. Dueholm M, Lundorf E, Hansen ES, et al. Accuracy of magnetic resonance imaging and transvaginal ultrasonography in the diagnosis, mapping, and measurement of uterine myomas. Am J Obstet Gynecol. 2002;186:409–15.
17. Clark TJ, Voit D, Gupta JK, Hyde C, Song F, Khan KS. Accuracy of hysteroscopy in the diagnosis of endometrial cancer and hyperplasia: a systematic quantitative review. JAMA. 2002;288:1610–21.
18. Marret H, Fauconnier A, Chabbert-Buffet N, et al. Clinical practice guidelines on menorrhagia: management of abnormal uterine bleeding before menopause. Eur J Obstet Gynecol Reprod Biol. 2010;152:133–7.
19. Dijkhuizen FP, De Vries LD, Mol BW, Brolmann HA, Peters HM, Moret E, et al. Comparison of transvaginal ultrasonography and saline infusion sonography for the detection of intracavitary abnormalities in premenopausal women. Ultrasound Obstet Gynecol. 2000;15:372–6.
20. Rodriguez GC, Yaqub N, King ME. A comparison of the Pipelle device and the Vabra aspirator as measured by endometrial denudation in hysterectomy specimens: the Pipelle device samples significantly less of the endometrial surface than the Vabra aspirator. Am J Obstet Gynecol. 1993;168:55–9.
21. Yarandi F, Izadi-Mood N, Eftekhar Z, et al. Diagnostic accuracy of dilatation and curettage for abnormal uterine bleeding. J Obstet Gynaecol Res. 2010;36:1049–52.
22. Goldstein SR. The role of transvaginal ultrasound or endometrial biopsy in the evaluation of the menopausal endometrium. Am J Obstet Gynecol. 2009;201(1):5–11.
23. Elsandabesee D, Greenwood P. The performance of Pipelle endometrial sampling in a dedicated postmenopausal bleeding clinic. J Obstet Gynecol. 2005;25:32–4.

24. Kelekci S, Kaya E, Alan M, Alan Y, Bilge U, Mollamahmutoglu L. Comparison of transvaginal sonography, saline infusion sonography, and office hysteroscopy in reproductive-aged women with or without abnormal uterine bleeding. Fertil Steril. 2005;84:682–6.
25. Valenzano MM, Lijoi D, Mistrangelo E, Fortunato T, Costantini S, Ragni N. The value of sonohysterography in detecting intracavitary benign abnormalities. Arch Gynecol Obstet. 2005;272:265–8.
26. Goldstein SR, Zeltser I, Horan CK, Snyder JR, Schwartz LB. Ultrasonography-based triage for perimenopausal patients with abnormal uterine bleeding. Am J Obstet Gynecol. 1997;177:102–8.
27. Klehr-Bathmann I, Kuhl H. Formation of ethinylestradiol in postmenopausal women during continuous treatment with a combination of estradiol, estriol and norethisterone acetate. Maturitas. 1995;21:245–50.
28. Kuhnz W, Heuner A, Hümpel M, et al. In vivo conversion of norethisterone and norethisterone acetate to ethinyl estradiol in postmenopausal women. Contraception. 1997;56:379–85.
29. Mansour D. Safer prescribing of therapeutic norethisterone for women at risk of venous thromboembolism. J Fam Plann Reprod Health Care. 2012;38:148–9.
30. Chu MC, Zhang X, Gentzschein E, et al. Formation of ethinyl estradiol in women during treatment with norethindrone acetate. J Clin Endocrinol Metab. 2007;92:2205–7.
31. Lidegaard Ø, Nielsen LH, Skovlund CW, et al. Risk of venous thromboembolism from use of oral contraceptives containing different progestogens and oestrogen doses: Danish cohort study, 2001-9. BMJ. 2011;343:d6423.
32. Kuhl H, Wiegratz I. Can 19-nortestosterone derivatives be aromatized in the liver of adult humans? Are there clinical implications? Climacteric. 2007;10:344–53.
33. Conard J, Plu-Bureau G, Bahi N, et al. Progestogen only contraception in women at high risk of venous thromboembolism. Contraception. 2004;70:437–41.
34. Gompel A, Carpentier S, Francès C, et al. Risk of venous thromboembolism and oral contraceptives. Lancet. 2002;359:1348–9.
35. Fulghesu AM, Magnini R, Piccaluga MP, Porru C. Ovulation induction in young girls with menometrorragia: a safe and effective treatment. Gynecol Endocrinol. 2014;30(2):117–20.
36. DeVore GR, Owens O, Kase N. Use of intravenous Premarin in the treatment of dysfunctional uterine bleeding – a double-blind randomized control study. Obstet Gynecol. 1982;59:285–91.
37. Munro MG, Mainor N, Basu R, Brisinger M, Barreda L. Oral medroxyprogesterone acetate and combination oral contraceptives for acute uterine bleeding: a randomized controlled trial. Obstet Gynecol. 2006;108:924–9.
38. American College of Obstetricians and Gynecologists. ACOG committee opinion no 557: management of acute abnormal uterine bleeding in nonpregnant reproductive-aged women. Obstet Gynecol. 2013;121:891–6.
39. De Francicis P, Cobellis L, Fornaro F, Sepe E, Torella M, Colacurci N. Low-dose hormone therapy in the perimenopause. Int J Gynaecol Obstet. 2007;98:138–42.
40. Kaunitz AM, Bissonnette F, Monteiro I, et al. Levonorgestrel-releasing intrauterine system for heavy menstrual bleeding improves hemoglobin and ferritin levels. Contraception. 2012;86:452–7.
41. Kaunitz AM, Inki P. The levonorgestrel-releasing intrauterine system in heavy menstrual bleeding: a benefit-risk review. Drugs. 2012;72:193–215.
42. Vilos GA, Marks J, Tureanu V, Abu-Rafea B, Vilos AG. The levonorgestrel intrauterine system is an effective treatment in selected obese women with abnormal uterine bleeding. J Minim Invasive Gynecol. 2011;18:75–80.
43. Lee NK, et al. Prognostic factors for uterine cancer in reproductive-aged women. Obstet Gynecol. 2007;109(3):655–62.
44. Penner KR, Dorigo O, Aoyama C, Ostrzega N, Balzer BL, Rao J, et al. Predictors of resolution of complex atypical hyperplasia or grade 1 endometrial adenocarcinoma in premenopausal women treated with progestin therapy. Gynecol Oncol. 2012;124:542–8.
45. Campagnoli C, Abba C, Ambroggio S, Brucato T, Pasanisi P. Life-style and metformin for the prevention of endometrial pathology in postmenopausal women. Gynecol Endocrinol. 2013;29:119–24.

46. Session DR, Kalli KR, Tummon IS, Damario MA, Dumesic DA. Treatment of atypical endometrial hyperplasia with an insulin-sensitizing agent. Gynecol Endocrinol. 2003;17: 405–7.
47. Tasci Y, Polat OG, Ozdogan S, et al. Comparison of the efficacy of micronized progesterone and lynestrenol in treatment of simple endometrial hyperplasia without atypia. Arch Gynecol Obstet. 2014;2:1.
48. Gallos ID, Shehmar M, Thangaratinam S, Papapostolou TK, Coomarasamy A, Gupta JK. Oral progestogens vs levonorgestrel-releasing intrauterine system for endometrial hyperplasia: a systematic review and metaanalysis. Am J Obstet Gynecol. 2010;203(6):547.
49. Mann M, Koller E, Murgo A, Malozowski S, Bacsanyi J, Leinung M. Glucocorticoidlike activity of megestrol. A summary of Food and Drug Administration experience and a review of the literature. Arch Intern Med. 1997;157(15):1651–6.
50. Schindler AE, Campagnoli C, Druckmann R, Huber J, Pasqualini JR, Schweppe KW, Thijssen JH. Classification and pharmacology of progestins. Maturitas. 2008;61(1–2):171–80.
51. Mesci-Haftaci S, Ankarali H, Yavuzcan A, Caglar M. Endometrial curettage in abnormal uterine bleeding and efficacy of progestins for control in cases of hyperplasia. Asian Pac J Cancer Prev. 2014;15(8):3737–40.
52. Meden-Vrtovec H, Hren-Bozic M. Glandular cystic hyperplasia of the endometrium in the perimenopausal years. Acta Eur Fertil. 1988;19(1):49–52.
53. Tasci Y, Polat OG, Ozdogan S, Karcaaltincaba D, Seckin L, Erkaya S. Comparison of the efficacy of micronized progesterone and lynestrenol in treatment of simple endometrial hyperplasia without atypia. Arch Gynecol Obstet. 2014;290(1):83–6.
54. El Behery MM, Saleh HS, Ibrahiem MA, Kamal EM, Kassem GA, El Sayed Mohamed M. Levonorgestrel-releasing intrauterine device versus dydrogesterone for management of endometrial hyperplasia without atypia. Reprod Sci. 2014. pii: 1933719114542014. [Epub ahead of print].
55. Gallos ID, Krishan P, Shehmar M, Ganesan R, Gupta JK. LNG-IUS versus oral progestogen treatment for endometrial hyperplasia: a long-term comparative cohort study. Hum Reprod. 2013;28(11):2966–71. Doi: 10.1093/humrep/det320. Epub 2013 Aug 23.
56. Abu Hashim H, Zayed A, Ghayaty E, El Rakhawy M. LNG-IUS treatment of non-atypical endometrial hyperplasia in perimenopausal women: a randomized controlled trial. J Gynecol Oncol. 2013;24(2):128–34.
57. Ismail MT, Fahmy DM, Elshmaa NS. Efficacy of levonorgestrel-releasing intrauterine system versus oral progestins in treatment of simple endometrial hyperplasia without atypia. Reprod Sci. 2013;20(1):45–50.
58. Ravibabu K, Palla J, Chintada GS. A study of efficacy of ormeloxifene in the pharmacological management of dysfunctional uterine bleeding. J Clin Diagn Res. 2013;7(11):2534–6.
59. Bouchard P. Current and future medical treatments for menometrorrhagia during the premenopause. Gynecol Endocrinol. 2011;27 Suppl 1:1120–5.
60. Lethaby A, Duckitt K, Farquhar C. Non-steroidal antiinflammatory drugs for heavy menstrual bleeding. Cochrane Database Syst Rev. 2013;1, CD000400.
61. Lethaby A, Farquhar C, Cooke I. Antifibrinolytics for heavy menstrual bleeding. Cochrane Database Syst Rev. 2000;(4). Art. No.: CD000249. DOI: 10.1002/14651858.CD000249.
62. Lukes AS, Moore KA, Muse KN, Gersten JK, Hecht BR, Edlund M, et al. Tranexamic acid treatment for heavy menstrual bleeding: a randomized controlled trial. Obstet Gynecol. 2010;116:865–75.
63. Berntorp E, Follrud C, Lethagen S. No increased risk of venous thrombosis in women taking tranexamic acid. Thromb Haemost. 2001;86:714–5.

Chapter 8
Progestogens in Contraception

Johannes Bitzer

1 Development and Classification of Protestogens in Contraception

The progestogens used for contraception, like all major steroid hormones are characterized biochemically by a C 21 carbon skeleton. Different progestogens have been developed, from the basic C21 skeleton for contraception. Ethinyl substitution of testosterone was found to result in an orally active compound (ethisterone). Removal of the carbon at the C-19 position of ethisterone changed ethisterone from an androgen to a progestin, resulting in the development of a class of progestins referred to as 19-nortestosterone derivatives. Included in this class are commonly used progestins such as norethindrone, norethindrone acetate, levonorgestrel and ethynodiol diacetate.

The first orally active progestogens used in combined hormonal contraceptives were norethynodrel and norethisterone.

The reason for the development of other progestogens was the observation that the early progestogens had androgenic properties leading to unwanted effects on lipids and the skin. The progestogens used in Combined Hormonal Contraceptives can be classified according to different criteria as follows:

J. Bitzer, M.D. (✉)
Department Obstetrics and Gynecology, University Hospital Basel, Basel, Switzerland
e-mail: johannes.bitzer@usb.ch

H.J.A. Carp (ed.), *Progestogens in Obstetrics and Gynecology*,
DOI 10.1007/978-3-319-14385-9_8

1.1 *The Time of Introduction into the Market*

It has become common use to apply this "historical" classification. According to this 1st, 2nd, 3rd and 4th generation combined hormonal contraceptives can be distinguished:

- 1st generation: Norethynodrel, Norethisteron Acetate, (NET, NETA).
- 2nd generation: Levonorgestrel (LNG).
- 3rd generation: Gestodene, Desogestrel, Norgestimate (GEST, DES, NGM).
- 4th generation: Drospirenone (DROSP).

Cyproterone Acetate (CPA) and Chlormadinone Acetate (CMA) have never been included in this categorisation as CPA containing pills were originally classified as drugs to treat hyperandroegenism in women who required contraception. CMA was only introduced in some countries and was and is not internationally available. The same was true for Dienogest (DNG) which was developed in Germany and is mainly used in Germany.

1.2 *Classification According to Molecular Structure*

The molecular structure gives an indirect indication about the biologic action of the steroid.

Different groups of progestogen can be distinguished. These are described in Fig. 8.1.

1.2.1 Derivatives of Testosterone

The basic molecule is testosterone. Due to elimination at the C 19 position the molecule becomes more progestogenic and the different variations are called C 19 Nortestosterone derivatives (see Fig. 8.2). The structure closest to testosterone can be seen in Norenthynodrel, Norethindrone, Norethisterone Acetate, Lynestrenol (Estranes). Further modification has lead to from the original estranes to gonanes like Levonorgestrel, Gestoden, Desogestrel, Norgestimate.

A special molecule in this context is Dienogest.

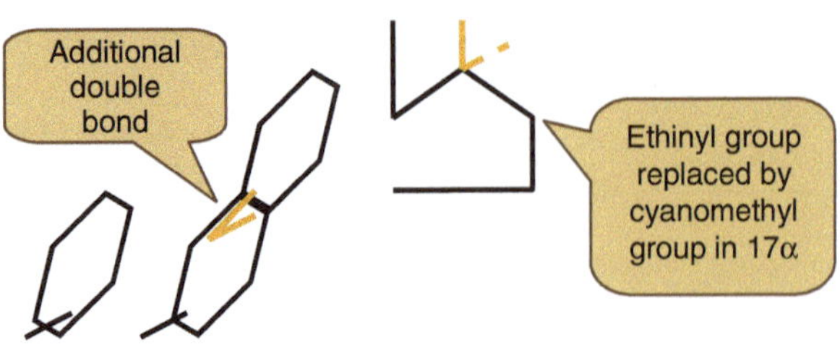

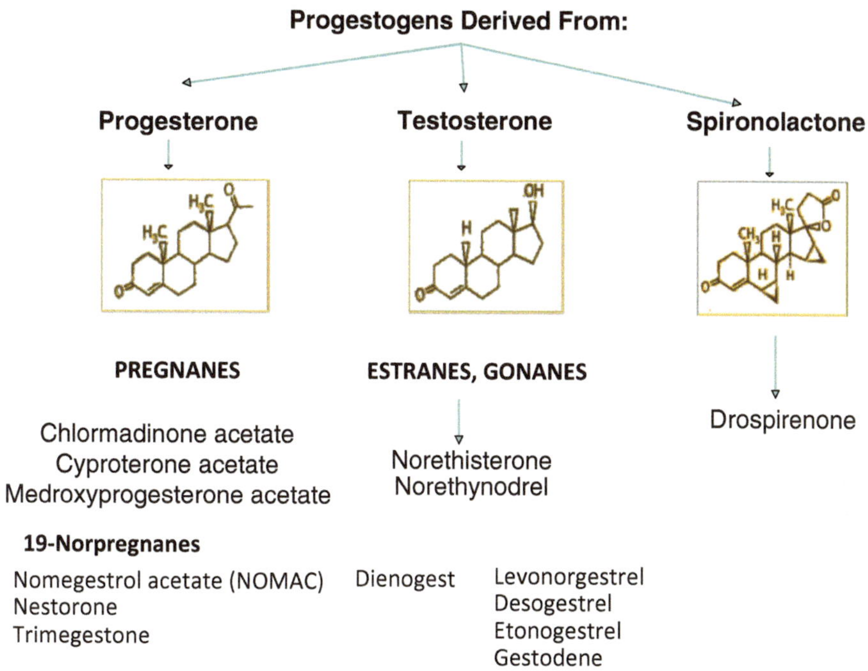

Fig. 8.1 The contraceptive toolbox

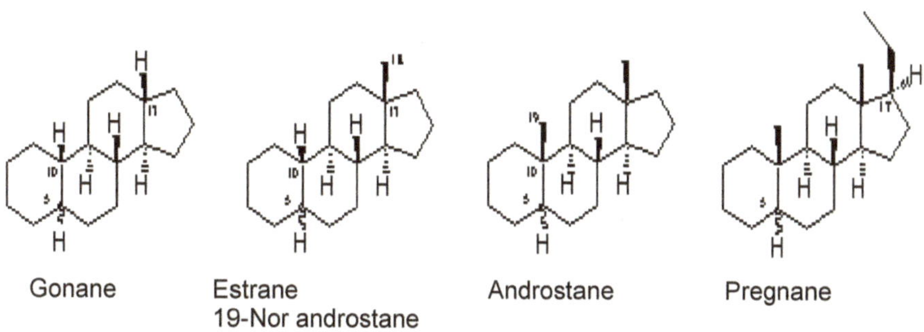

Fig. 8.2 C 19 Nortestosterone derivatives

Dienogest is a 19 Nortestosterone derivative with a Cyanomethyl group instead of the usual Ethinyl Group in the 17 alpha position.

1.2.2 Derivatives of Progesterone

Subgroups can be distinguished depending on the different positions of double bindings between C atoms and the type of C group added.

An important molecule is Medroxyprogesteronacetate.

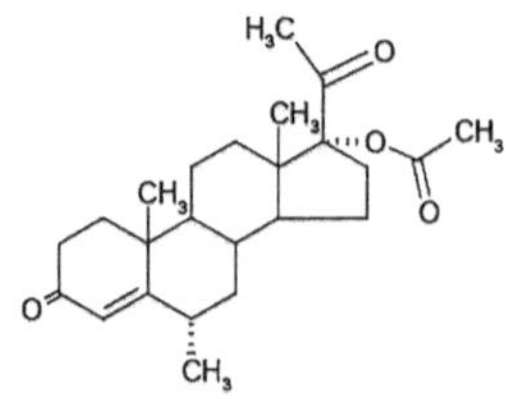

Another important progesterone derivatives in COCs are Cyproterone Acetate and Chlormadinone Acetate.

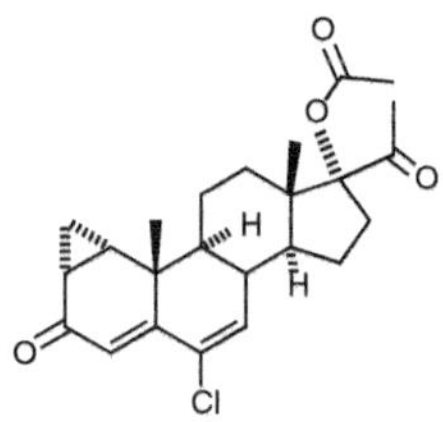

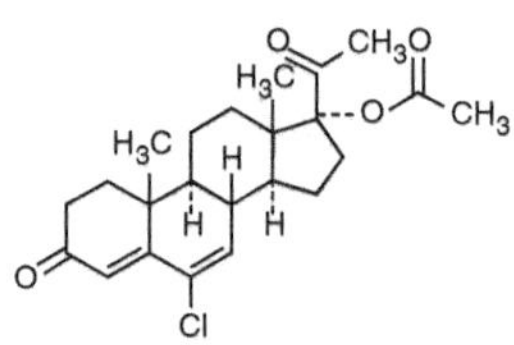

6-Chloro-3,20-dioxopregna-4,6-dien-17-yl acetate

Both have a C-21 structure and are basically different from Testosterone derivates in their action in the body (see below). Both orally active progestogens suitable for use in combined hormonal contraceptives.

Another subgroup of progesterone derivatives are the 19 Norpregnanes.

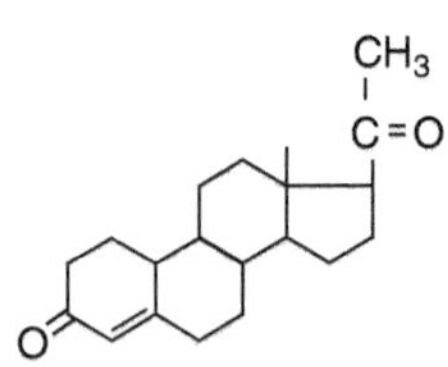

19-Norprogesterone

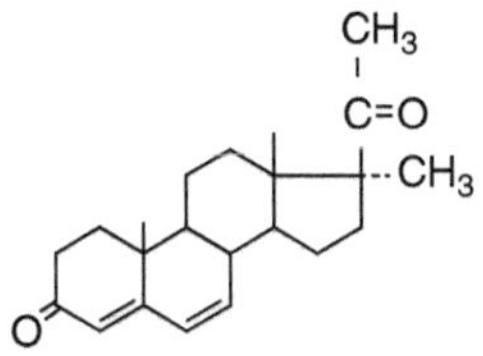

Demegestone

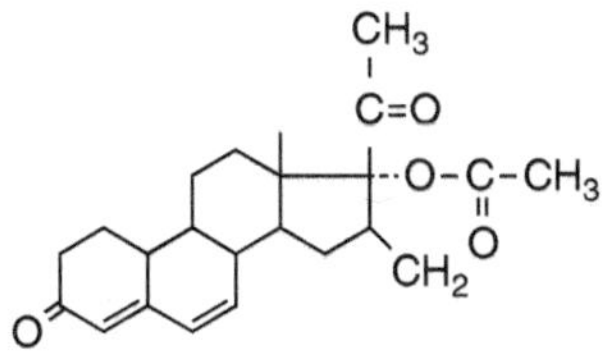

Nestorone

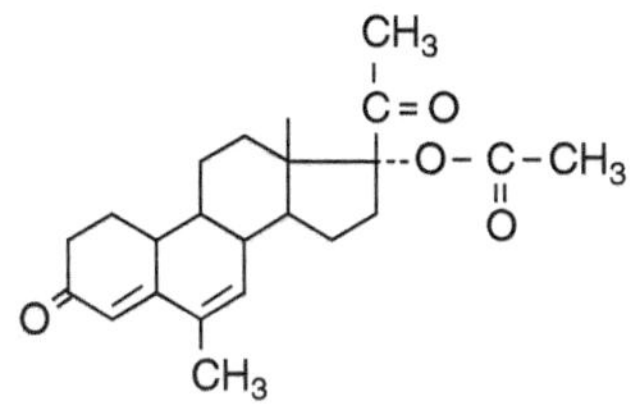

Nomegestrol acetate

Their structure is very similar to progesterone. At position 19 the C methyl group is removed.

They have strong progestogenic activity. Nomegestrol acetate is orally active.

1.2.3 Derivatives of Spironolactone

Drospirenone is the only progestogen which is derived from spironolactone, which is known to counteract sodium retention and has an antiandrogenic action on the skin.

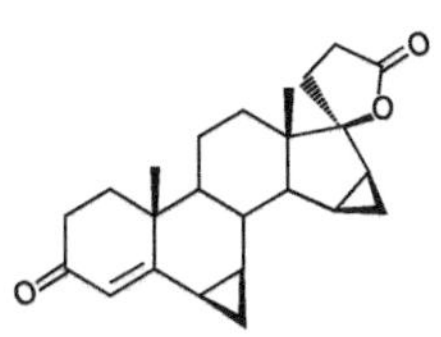

Drospirenon

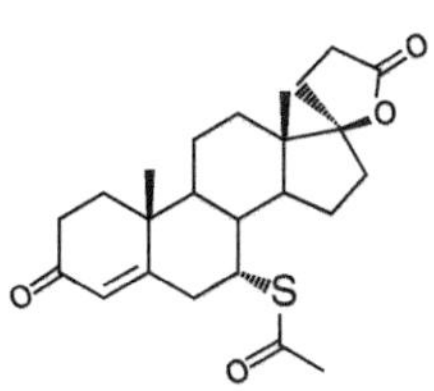

Spironolacton

The drospirenone Molecule is a "mixture" of the progesterone and the spironolactone molecule.

1.3 Classification According to Interaction with Steroid Receptors

Depending on structure progestogens have different interactions with the various steroid receptors in the body. Steroid receptors are located on the membrane of target cells and are linked to the DNA/RNA and Protein production via different messenger systems. Two properties can be distinguished:

(a) The binding capacity, i.e., the intensity with which the ligand of the molecule binds to appropriate receptor sites.
(b) The direction and the intensity of the induced action. The steroid can bind to the receptor but does not induce any activity. The final effect is a reduction of the activity mediated by the receptor by competitive inhibition (the steroid binds to the receptor and thus blocks the binding site for a steroid that would activate receptor activity).

The receptors to which progestogens bind can be the following:

- Progestogen receptor: The most important receptor to induce the desired effect.
- Androgen receptor: Activation of androgen receptors mediatc androgenic effects on hair growth and activity of the sebaceous glands. Some progestogens bind to this receptor and can either block or activate it (antiandrogenic properties see below).
- Estrogen receptor: This receptor mediates effects in many tissues especially in the endometrial cells.
- Glucocorticoid receptor: The glucocorticoid effect is linked to the activation of the coagulation system.
- Mineralocorticoid receptor: This receptor mediates sodium retention.

Based on this classification of receptor activities several groups of progestogens can be differentiated. The clinical consequences of these differences are still controversial and research is going on.

- Androgenic progestogens: Norethyodrel, Levnorgestrel and Norgestimate have androgenic properties. LNG seems to have not only an androgenic but also an antiestrogenic effect. Antiandrogenic progestogens: The most important compounds are the progesterone derivatives CPA and CMA and drospirenone. These molecules bind to the receptor and exert competitive inhibition.
- Mildly antiandorgenic or neutral progestogens: GEST, DES, mainly interact with the progestogene receptor alone. Dienogest has a weak antiandrogenic action.
- Antimineralocorticoid progestogens: Only Drospirenone has antimineralocorticoid action.

The different receptor effects can be described are shown in Table 8.1.

2 Health Risks of Combined Hormonal Contraceptives in Relation to the Progestogen

2.1 Third Generation

Several registry based studies published in the British Medical Journal (BMJ), particularly the Danish registry indicated that there might be an increased risk of venous thromboembolism (VTE) associated with the intake of 3rd and 4th

Table 8.1 Contraceptive progestogens and extra-progestogenic effects (adapted from Benagiano et al. [24]

Glucocorticoid activity	Estrogenic activity	Antiestrogenic activity	Androgenic activity	Antiandrogenic activity	Mineralocorticoid activity
Medroxy progesterone acetate	Norethisterone acetate	Levonorgestrel	Levonorgestrel	Cyproterone acetate	Drospirenone
Megestrol acetate			Desogestrel	Chlormadinone acetate	
			Gestodene	Dienogest	

generation COCs compared to preparations containing the progestogen levonorgestrel (LNG) or other first and second generation progestogens [1–5].

The relative risk increase was approximately 2 and absolute attributable risk was given—dependent on the base prevalence rate—between 2 and 8 per 10,000 users [6]. These results contrast with those of published prospective cohort studies, sponsored by Bayer Health Care, at the request of health authorities for a large postmarketing survey, which did not find a differences between the various generations of progestogens The discrepancy led to intensive scientific discussion among epidemiologists about possible confounders and biases in the published studies [7, 8]. The present recommendation is to inform women about the controversial results and the suspected, but not confirmed risk of the newer preparations; Health care professionals are required to balance the risks and benefits of the different preparations in a process of shared decision making with the individual woman. Women should be advised to continue hormonal contraception, in order to avoid the previously observed rise in the number of artificial abortions following the "pill scare" in 1995 [9, 10].

The unresolved scientific debate about VTE has continued, and was given new impetus by another publication based on the Danish Registry indicating that treatment with the transdermal patch, the vaginal ring and the implant were also associated with an increased risk of VTE, whereas the LNG releasing intrauterine system (LNG-IUS) did not or even decreased the VTE risk compared to non users. In this publication the authors recommended and gave practical advice how to switch from the contraceptives mentioned to either a LNG containing pill, a LNG-IUS or a non hormonal method [11, 12].

These publications with warnings about the increased risk of 3rd and 4th generation contraceptives lead to an intensive debate among epidemiologists about the limitations of registry based observational studies in comparison to other study designs and about the clinical validity of these results [13–17].

In summary, most studies suggest that the third generation progestins, desogestrel and gestodene (but not norgestimate), may be associated with a higher risk of venous thromboembolism when compared with levonorgestrel, a second generation progestin. However, the absolute excess risk is small, and may be outweighed by the many benefits of OCs, including prevention of an unwanted pregnancy, and a reduction in ovarian and endometrial cancer risk.

2.2 Antiandrogens

There are some special concerns and considerations regarding progestogens with an antiandrogenic action.

Cyproterone acetate. In two studies, a higher risk of VTE was seen when compared with contraceptives containing levonorgestrel [18, 19]. In a report from the Danish National Registry the risk was not significantly different from

LNG (absolute risk 4.2 and 3.1 per 10,000 woman-years for levonorgestrel and cyproterone acetate, respectively) [20].

Chlormadinone Acetate. There are no large epidemiological studies with CMA, but the post marketing studies especially in Germany and Austria did not show an increased risk for VTE in users. The European Authorities demand however more surveillance studies to be able to determine the cardiovascular risk of chlormadinone which has been shown to be effective in reducing acne and dysmenorrhea.

Drospirenone. Drospirenone, a progestin that also has antiandrogen and antimineralocorticoid properties, has been associated with a greater risk of VTE when compared with levonorgestrel in some, but not all studies (see also above).

Two observational studies have reported that oral contraceptives containing drospirenone were associated with an excess risk of venous thromboembolism (similar in magnitude to the third generation progestins) [1, 2]. Two previous large, prospective, surveillance studies of new users of drospirenone-containing oral contraceptives subsequently reported that the thromboembolism risk was no different from that for other OCs It has been estimated that 9,000 women would need to be treated with a drospirenone OC in order to see one additional case of venous thromboembolism [7, 8]. After the publication of the two surveillance studies, two additional case-control studies and a registry-based cohort study reported a two- to threefold increased risk of VTE with OCs containing drospirenone compared with levonorgestrel [3–5].

An FDA sponsored study published after the Drug Safety Communication utilized computerized data files from two integrated medical care programs and two state Medicaid programs to obtain data regarding the risk of several cardiovascular endpoints in combined hormonal contraceptives users. The authors identified a final cohort that included 189,210 person-years of exposure to drospirenone. In adjusted analyses, drospirenone use was associated with a significantly higher risk of VTE relative to low-estrogen comparators (RR 1.74; 95 % CI 1.42–2.14) [21].

In 2012, based upon available data, the US Food and Drug Administration (FDA) added revised labeling to all oral contraceptives containing drospirenone, stating that they may be associated with up to a threefold higher risk of VTE compared with OCs with levonorgestrel and some other progestins [22]. The FDA does not advise women to stop drospirenone-containing OCs, but does suggest that an individual's risk of VTE be assessed before starting one in a new OC user, or before considering using one in a woman who has been on an OC not containing drospirenone. Lastly, the warning notes that the VTE risk with drospirenone is small and still lower than the risk of VTE during pregnancy.

In 2011, the European Medicines Agency also concluded that drospirenone-containing birth control pills carry a higher risk of venous thromboembolism, but noted the overall risk of blood clot from any birth control method remains small and stopped short of advising women to stop taking pills containing drospirenone [23].

3 Progestogen Only Contraception (Short Overview)

3.1 Common Features of Progestogen Only Contraception

3.1.1 General Principle of Action

The basis for progestogen only contraception lies in the specific action of progestogens on reproductive physiology. These actions are type and dose dependent and include:

- Inhibition of ovulation.
- Change of the cervical mucus to make it impenetrable.
- Endometrial changes which make implantation either difficult or impossible.
- Changes in tubal mobility.

The common clinical important features of these methods are:

3.1.2 Very Low or Absence of Cardiovascular Risks

Progestogens have very little impact on the coagulation system (see above). Their effects on blood flow and contractility of vessel walls is very limited. Epidemiological studies do not show any significant risk for thromboembolic venous or arterial disease. Therefore progestogen only contraceptives can be used in women who have a contraindication for combined hormonal contraceptives (WHO MEC Category 4) or where the use is not advised (Category 3).

Contraindications to combined hormonal contraception include:

- Women postpartum (during the first 21 days, lactating women).
- Women with a combination of cardiovascular risk factors (obesity, smoking, age).
- Women with specific venous thromboembolic risk factors (thrombophilia, family history, longer immobilisation, acute DVT with anticoagulant therapy).
- With women with specific arterial risk factors (hypertension, migraine with aura, valvular heart disease, past ischemic heart disease).

3.1.3 Contraindications for Progesterone Only Contraception

There are few contraindications to progestogen only contraception, but these include: Breast Cancer, active liver disease, benign and malignant liver tumours (except nodular hyperplasia).

3.1.4 Side Effects

The most frequent side effects attributed to the action of progestogens are acne, mild hirsutism, depressive mood, sexual pain and weight gain. This is however not based on prospective clinical comparative trials but mainly observational data. The most frequent side effect of continuous use of progestogens is irregular bleeding.

3.1.5 Additional Benefits and Therapeutic Indications

Progestogen only contraception confers several important additional benefits:

- Lactation: The progestogen only contraceptives can be used in lactating women because there is no reduction in milk production and no negative effect on the newborn.
- Menstrual symptoms: Contraceptive progestogens can, due to their antimitotic and transformational action on the endometrial cells reduce the frequency and intensity of uterine bleeding. The contraceptive progestogens with ovulation inhibition can reduce dysmenorrhea. Additionally progestogens block the synthesis of prostaglandins in the endometrium by reducing the endometrial thickness.
- Menstrual Migraine: Progestogens in continuous use reduce the intensity of menstrual migraine.
- Endometriosis: Progestogens can reduce the proliferative activity of the endometrium.

3.1.6 High Efficacy in Typical Use

The long acting preparations of contraceptive progestogens (injections, implants, medicated IUDs) have a very high efficacy in typical use due to the fact that the contraceptive effect is largely independent of the compliance of the user.

3.2 Routes of Application

Progestogen only contraception can be applied orally, by injection, as a intrauterine system or as implant.

3.2.1 Oral Preparations

Compositions

Older oral preparations containing Levonorgestrel and Norethisterone are unable to inhibit ovulation but their contraceptive effect is based on the action on the cervical mucus which becomes impenetrable to sperm. An additional effect of these progestogen only pills is effect on the endometrium desynchronising ovulation and endometrium transformation and preparation for implantation.

These preparations should be taken at more or less the same time every day in order to exert the contraceptive effect. The typical user Pearl Index is between 6 and 8. The newer progestogen only pill with 75 μg of desogestrel daily is taken continuously without a 7 day break. 75 μg of desogestrel inhibits ovulation inhibition and is as effective as combined hormonal contraceptives. This pill is also called an estrogen free inhibitor of ovulation. No major health risks are known.

Contraindications

Breast Cancer, active liver disease, benign and malignant liver tumours (except nodular hyperplasia).

Side Effects

Due to the daily intake needed for ovulation suppression there is no phase of progestogen withdrawal which is the reason for the bleeding occurring during the pill free interval during the use of combined hormonal contraceptives in the 21/7 regimen. Irregular bleeding is therefore the main complaint which may lead to discontinuation. Other progestogenic side effects such as acne, weight gain, depressed mood are rare.

Special Benefits

There is alleviation of menstrual migraine, pain reduction in patients with endometriosis and reduction of hypermenorrhea and dysmenorrhea.

3.2.2 Progestogen Implants

Compostions, Type of Preparations

Hormonal implants are subdermally inserted contraceptives that provide reliable contraception for 3–5 years. The matrices are inert or biologically degradable rods or capsules which release the respective steroid continuously over a lengthy period of time. The Population Council in New York has studied long-term contraception with subdermal hormonal implants since 1966. The hormone implants consist of one or several small flexible rods or a capsule inserted under the skin of the upper arm. Depending on the product, they release the progestins megestrol acetate, norethindrone, norgestrinone or etonogestrel for a period of 1 year to 5 years.

Norplant® was composed of six rods. Each rod contains 36 mg of levonorgestrel. The total duration of action of these 6 rods was 5 years. The product has not been marketed since 2002.

Norplant II® (Jadelle®), Norplant®'s successor product, is composed of two flexible silicone rods (43 mm Å–2.5 mm) each containing 75 mg of levonorgestrel and also has a duration of action of 5 years. The same product is commercially available in China under the name Sinoplant.

The etonogestrel Releasing Hormonal Implant Implanon®

Composition: Implanon® is an etonogestrel-releasing hormonal implant. The rod is 4 cm long and 2 mm in diameter and is composed of 40 % ethylene vinyl acetate (EVA) and 60 % (68 mg) etonogestrel (3-keto-desogestrel). The duration of action after subdermal implantation is 3 years.

Mechanism of Action: The main effect is ovulation inhibition, although this inhibitory effect is less at the end of the 3 years. Addtional contraceptive effects are the change in the composition of the cervical mucus and making the endometrium less receptive for a theoretical mplantation. The PI is 0.38 pregnancies per 100 women-years of use, which is similar to that of other long-acting methods of contraception.

The concentration falls over time at a rate that depends on body weight. Clinical experience with Implanon® in women weighing more than 80 kg is limited. Available data do not show a decrease in efficacy in obese women.

Accurate placement is crucial to the product's reliability. There are reports of incorrect insertion of the Implanon® rod, possibly making the contraceptive rod impossible to palpate and difficult to find. To make the product easier to use safely and simpler to locate, the system was upgraded with Implanon NXT®. Efficacy may be hampered by drugs affecting the metabolism of etonogestrel like antiviral drugs. No major health risks are known. There is no concern regarding bone loss. Contraindications include Breast Cancer, active liver disease, benign and malignant liver tumours (except nodular hyperplasia).

Side Effects: As with other progesterone only contraceptives, the most frequent side effect are unscheduled bleeding. These lead to various degrees of discontinuation, (approximately 15–18 % in USA and Europe and 3–4 % in Southeast Asia and Russia. In approximately 5 % of users the following side effects were reported: acne, headache, weight gain, mastalgia, vaginal infections and bleeding disorders. Interactions with broad-spectrum antibiotics, St. John's wort, a number of antiepileptic agents and mood-altering drugs have been documented. It should be borne in mind that placement and removal require special training.

Additional Benefits: Clinical studies have shown that Implanon® is also effective in treating heavy dysmenorrhea. However, the product is not approved for that purpose. Its use in this therapeutic indication is an off-label use Another possible beneficial effect is the diminuition of pain caused by endometriosis.

3.2.3 Injections (Intramuscular and Subcutaneous)

Composition

Medroxy progesterone acetate (DMPA) is available for injection as a depot either. In a dose of: 150 mg/1 mL injected intramuscularly or in a dose of 104 mg/0.65 mL injected subcutaneously. Both preparations last for 3 months. The mechanism of action of DMPA is similar to other progestogens and includes a strong antigonadotropic effect, inhibition of ovulation, inhibition of endometrial proliferation, and changes in the cervical mucus making the mucus impenetrable for sperm. The dose is a standard dose, and no adjustment is necessary for body weight. There is no reduction in efficacy with concurrent medication.

Health Risks

No major health risks have been found. However, long term DMPA use induces an unwanted increase of LDL Cholesterol and reduces peripheral arterial flow-mediated dilatation, which are matters of concern. DMPA does not alter coagulation factors nor increase blood pressure. Some studies have found an increased VTE risk but these studies have methodological weaknesses and a small number of cases. Nonetheless the World Health Organization has attributed category 23 to DMPA in women with current VTE, a history of stroke or ischemic heart disease.

The primary concern is the effect of DMPA on bone density. DMPA suppresses endogenous estrogen production from the ovaries by its strong antigonadotropic action. Compared to nonusers, the bone mineral density at the hip and spine of DMPA users decreases by 0.5–3.5 % after 1 year and 5.7–7.5 % after 2 years of use. There is therefore concerns about the use of DMPA in adolescents when the accumulation of bone mass is at its peak, and in premenopausal women as there may be an increased rate of bone density loss. However, many studies have shown that the bone loss is reversible and the best evidence available at present indicates that that DMPA does not reduce peak bone mass and does not increase the risk of osteoporotic fractures in later life in women with an average risk of osteoporosis.

At present, DMPA is considered to be contraindicated when pregnancy is planned within the next year, in the presence of osteoporosis and known risk factors for fractures, or in the presence of hypothalamic amenorrhea, anorexia nervosa or chronic glucocorticoid therapy.

Side Effects

Side effects include menstrual irregularities (during the first 3–6 months irregular bleeding and spotting, later there may be amenorrhea in up to 75 % of users). There may be weight gain of between 3 and 6 kg (especially in young obese women). Headache, abdominal discomfort and pain, dizziness, nervousness and asthenia have also been described.

Benefits

The benefits include, reduction of heavy menstrual bleeding due to the high incidence of amenorrhea after longer use, a reduced risk of pelvic inflammatory disease, reduction of endometriotic pain, fewer painful crisis in women with sickle cell disease and reduction in vasomotor symptoms.

3.2.4 Levonorgestrel Containing Intrauterine Systems

There are two types of Levonorgestrel containing intrauterine systems (a) Levonorgestrel-releasing Intra-uterine device IUD (LNg 20)—The LNg20 IUD consists of a T-shaped polyethylene frame with a collar containing 52 mg of levonorgestrel dispersed in polydimethylsiloxane attached to a vertical stem. In vivo, the LNg20 IUD initially releases 20 mcg of levonorgestrel daily, declining to 10–14 mcg per day after 5 years. The LNg20 IUD should be replaced after 5 years (b) Levonorgestrel-releasing IUD (LNg14) It contains 13.5 mg levonorgestrel, which is initially released at a rate of 14 mcg daily, declining to 5 mcg/day after 3 years. It is smaller in size (T shape 28 by 30 mm) than the LNg20 (T shape 32 by 32 mm) and has a smaller inserter diameter (3.8 mm versus 4.75 mm). It contains a silver ring to distinguish it on ultrasound from other IUDs and make it detectable by X-ray. The LNg14 IUD should be replaced after 3 years.

Both types of Levonorgestrel-releasing IUD's act by thickening the cervical mucus, making it impervious to sperm, by causing endometrial decidualization and glandular atrophy, inducing expression of glycodelin A in endometrial glands which inhibits the binding of sperm to the ovum and in partial inhibition of ovulation (in approx. 20 % of users).

Both types of Levonorgestrel-releasing IUDs are efficacious. The cumulative pregnancy rate is 0.5–1.1 after 5 years of continuous use with the LNG 20 IUD. The 3 year cumulative pregnancy rate is 0.9 with the LNG 14 IUD.

Contraindications

Contraindications include severe deformity of the uterine cavity, acute sexually transmitted infections, unexplained vaginal bleeding, current breast cancer, Wilson's disease and known or suspected pregnancy.

Side Effects

The major side effect is irregular bleeding, which is very common during the first 3–6 months. At 24 months 50 % of LNG20 users have amenorrhea, 25 % have oligomenorrhea and 11 % have spotting. The pattern is similar with the LNG 14 IUD with less amenorrhea (13 versus 24 % after 3 years).

Other side effects are rare and include breast tenderness, mood changes and acne.

Additional Benefits

There are many benefits to using the Levonorgestrel-releasing IUD. The major benefit is reduction in heavy menstrual bleeding and dysmenorrhea in patients without organic pathology and bleeding due to bleeding diathesis including anticoagulation therapy. The efficacy regarding reduction of bleeding intensity in women with fibroids and adenomyosis is yet unclear and under investigation. The Levonorgestrel-releasing IUD protects from pelvic inflammatory disease, due to cervical mucus thickening which acts as a barrier towards ascending infections. The Levonorgestrel-releasing IUD can be used to treat endometriosis. There is endometrial protection in premenopausal and menopausal women using estrogen replacement, and a concomitant reduction of the risk of endometrial cancer. The Levonorgestrel-releasing IUD can also be used to treat endometrial hyperplasia and cancer.

Special Consideration

LNG IUDs and especially the LNG14 IUD are a highly effective method of contraception in women under age 20 due to the facilitation of compliance and the high continuation rate. Rates of infection are similar in adolescent users to those of adults. Nulliparity is not a contraindication for the use of this method of contraception.

References

1. Lidegaard O, Lokkegaard E, Svendsen AL, Agger C. Hormonal contraception and risk of venous thromboembolism: national follow-up study. BMJ. 2009;339:b2890.
2. van Hylckama Vlieg A, Helmerhorst FM, Vandenbroucke JP, Doggen CJ, Rosendaal FR. The venous thrombotic risk of oral contraceptives, effects of estrogen dose and progestagen type: results of the MEGA case-control study. BMJ. 2009;339:b2921.
3. Jick SS, Hernandez RK. Risk of non-fatal venous thromboembolism in women using oral contraceptives containing drospirenone compared with women using oral contraceptives containing levonorgestrel: case-control study using United States claims data. BMJ. 2011;340:d2151.
4. Parkin L, Sharples K, Hernandez RK, Jick SS. Risk of venous thromboembolism in users of oral contraceptives containing drospirenone or levonorgestrel: nested case-control study based on UK General Practice Research Database. BMJ. 2011;340:d2139.
5. Lidegaard Ø, Nielsen LH, Skovlund CW, Skjeldestad FE, Løkkegaard E. Risk of venous thromboembolism from use of oral contraceptives containing different progestogens and estrogen doses: Danish cohort study 2001–9. BMJ. 2011;343:d6423.
6. Heinemann LAJ, Dinger JC. Range of published estimates of venous thromboembolism incidence in young women. Contraception. 2007;75:328–36.

7. Dinger JC, Heinemann LAJ, Kuhl-Habich D. The safety of a drospirenone-containing oral contraceptive: final results from the European Active Surveillance study on Oral Contraceptives based on 142,475 women-years of observation. Contraception. 2007;75:344–54.
8. Seeger JD, Loughlin J, Eng PM, et al. Risk of thromboembolism in women taking ethinylestradiol/drospirenone and other oral contraceptives. Obstet Gynecol. 2007;110:587–93.
9. Skjeldestad FE. Increased number of induced abortions in Norway after media coverage of adverse vascular events from the use of third-generation oral contraceptives. Contraception. 1997;55:11–4.
10. Goodyear-Smith A, Arroll B. Termination of pregnancy following panic-stopping of oral contraceptives. Contraception. 2002;66:163–7.
11. Lidegaard O, Nielsen LH, Skovlund CW, Lokkegaard E. Venous thrombosis in users of non-oral hormonal contraception: follow-up study, Denmark 2001–10. BMJ. 2012;344:e2990.
12. Lidegaard O, Milsom I, Geirssons RT, Skjeldestrad FE. Hormonal contraception and venous thromboembolism. Acta Obstetricia et Gynecologica Scandinavica. 2012;91:769–78.
13. Spitzer WO. The aftermath of a pill scare: regression to reassurance. Hum Reprod Update. 1999;5:736–45.
14. Lewis MA, MacRae KD, Kuhl-Habichl D, Bruppacher R, Heinemann LA, Spitzer WO. The differential risk of oral contraceptives: the impact of full exposure history. Hum Reprod. 1999;14:1493–9.
15. Dinger J. Oral contraceptives and venous thromboembolism: old questions revisited. J Fam Plann Reprod Health Care. 2009;35:211–3.
16. Shapiro S, Dinger J. Risk of venous thromboembolism among users of oral contraceptives: a review of two recently published studies. J Fam Plann Reprod Health Care. 2010;36(1):33–8.
17. Heinemann K, Heinemann LA. Comparative risks of venous thromboembolism among users of oral contraceptives containing drospirenone and levonorgestrel. J Fam Plann Reprod Health Care. 2011;37:132–5.
18. Seaman HE, de Vries CS, Farmer RD. The risk of venous thromboembolism in women prescribed cyproterone acetate in combination with ethinyl estradiol: a nested cohort analysis and case-control study. Hum Reprod. 2003;18:522.
19. Vasilakis-Scaramozza C, Jick H. Risk of venous thromboembolism with cyproterone or levonorgestrel contraceptives. Lancet. 2001;358:1427.
20. Lidegaard Ø. Absolute and attributable risk of venous thromboembolism in women on combined cyproterone acetate and ethinylestradiol. J Obstet Gynaecol Can. 2003;25:575.
21. FDA Office of Surveillance and Epidemiology. Combined hormonal contraceptives (CHCs) and the risk of cardiovascular disease endpoints. http://www.fda.gov/downloads/Drugs/DrugSafety/UCM277384.pdf. Accessed 3 Feb 2012.
22. US Food and Drug Administration. Birth control pills containing drospirenone: label change-products may be associated with a higher risk for blood clots. http://www.fda.gov/Safety/MedWatch/SafetyInformation/SafetyAlertsforHumanMedicalProducts/ucm299605.htm. Accessed 17 Apr 2012.
23. European Medicines Agency. PhVWP monthly report on safety concerns, guidelines and general matters. 2012; 1201. http://www.ema.europa.eu/docs/en_GB/document_library/Report/2012/01/WC500121387.pdf. Accessed 17 Apr 2012.
24. Benagiano G, Carrara S, Filippi V. Safety, efficacy and patient satisfaction with continuous daily administration of levonorgestrel/ethinylestradiol oral contraceptives. Patient Prefer Adherence. 2009;3:131–43.

Chapter 9
Progestogens and Endometriosis

Howard J.A. Carp, David Soriano, and Matityahu Zolti

1 Introduction

Endometriosis is a chronic and recurrent condition in which endometrial-like tissue is found outside the uterus. Endometriotic lesions may be superficial, deeply infiltrating or ovarian. Superficial lesions appear as black, dark brown or bluish coloured lesions, or as small cysts, covering old hemorrhages, surrounded by various degrees of fibrosis. Lesions are mostly located in the peritoneum, pouch of Douglas, ovaries or sacroiliac ligaments. Endometriosis is defined as deeply infiltrating if penetrating more than 5 mm under the peritoneum [1]. Infiltrating lesions may invade adjacent organs (bladder, bowel or rectum) leading to a wide range of symptoms. Endometriotic lesions may be found in organs as distant as liver, lung or brain. In the ovaries, endometriosis may form endometriomas, with smooth walled cysts filled with a "chocolate" like material. Some women with endometriosis experience extreme pain and/or infertility, while others have less or minor symptoms, or the symptoms appear late in the course of the disease. Endometriotic lesions are surrounded by an inflammatory reaction which may lead to adhesions ranging from filmy synechia to dense adhesions which obliterate all planes of separation between the various organs involved.

H.J.A. Carp, M.B.B.S., F.R.C.O.G. (✉)
Department of Obstetrics & Gynecology, Sheba Medical Center, Tel Hashomer 52621, Israel

Sackler School of Medicine, Tel Aviv University, Tel Aviv, Israel
e-mail: carp@netvision.net.il

D. Soriano, M.D.
Sackler School of Medicine, Tel Aviv University, Tel Aviv, Israel

Endometriosis center, Sheba Medical Center, Tel Hashomer, Israel

M. Zolti, M.D.
Obstetrics and Gynecology, Sheba Medical Center, Tel Hashomer, Israel

H.J.A. Carp (ed.), *Progestogens in Obstetrics and Gynecology*,
DOI 10.1007/978-3-319-14385-9_9

The exact prevalence is unknown but estimates range from 5 to 10 % of women of reproductive age, to 50 % of infertile women [2, 3]. The combination of laparoscopy and histological confirmation is the gold standard for diagnosis of the disease. However, the condition may be suspected by the symptoms, signs, gynecological examination, and imaging techniques. Invasive surgical techniques are not always essential for managing the disease. However, in cases of severe symptoms and findings it can offer a relief for long time. Diagnosis, although less accurate, can be obtained by non-invasive methods such as transvaginal sonography which consider today as the first method for evaluation or magnetic resonance imaging. Ovarian endomeriomas may be diagnosed by either imaging technique, Adhesions can be diagnosed by transvaginal ultrasound, with the sliding sign technique. This technique assesses whether the uterus and ovaries move freely over the adjacent organs and tissues. Magnetic resonance imaging can be useful in diagnosing all locations of endometriosis. A detailed non-invasive diagnosis of extension within the pelvis can facilitate the choice of a safe and adequate surgical or medical approach. Treatment is dependent on the patient's age and needs. Treatment may be required to reduce pain, enhance fertility, or both or to prevent recurrence. However, except in cases of a solitary nodule which is completely resected or focal lesions such as endometriosis, in scar, complete resolution is not possible as yet. Presently all treatment modalities provide temporary relief. This chapter describes the features of endometriosis which are relevant to progestogen treatment. It is not intended to be a full treatise on endometriosis.

2 Inflammatory Reaction Around Endometriotic Deposits

As long ago as 1922, it was recognized that neoangiogenesis, fibrosis and hemosiderin accumulation were intrinsic microscopic characteristics of the endometriotic lesion. In endometriosis, peritoneal macrophages are activated, and increased cytokine production is found. However, the macrophages cannot carry out phagocytosis of the ectopic endometrium [4]. The ectopic endomerium has been reported to contain a protein (Endo 1) [5] which binds to peritoneal macrophages, increasing their production of interleukin 6, and reducing their phagocytic ability [6]. Hence, there is compromised immune surveillance in endometriosis. There is also evidence of compromised innate immunity and natural killer cell activity in the peritoneal fluid in endometriosis [7]. In endometriosis, the peritoneal fluid has high concentrations of cytokines, growth factors, and angiogenic factors, derived from the lesions themselves, the secretory products of macrophages and other immune cells. Once endometriotic lesions are formed, they secrete several proinflammatory cytokines which recruit macrophages and T lymphocytes in to the peritoneum such as IL-1, IL-8, TNF-α, IFN-γ. Several angiogenic factors are also expressed by endometriotic lesions such as IL-1, IL-6, IL-8, EGF, Fibroblast growth factor, insulin-like growth factor, vascular endothelial growth factor (VEGF), and Endo I. thus explaining the angiogenesis in the peritoneal cavity [8]. Figure 9.1 shows the intense inflammatory

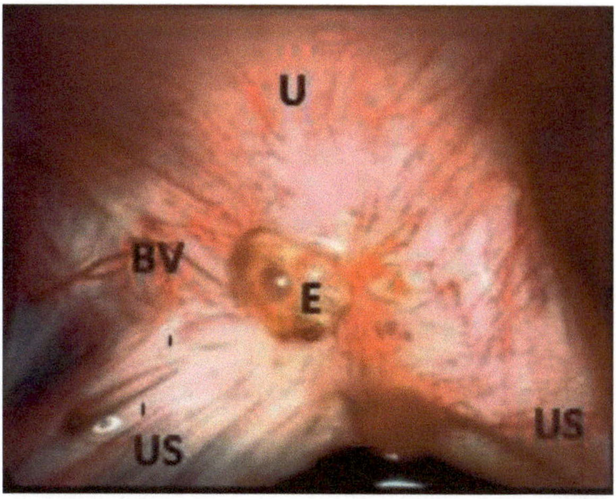

Fig. 9.1 Inflammatory reaction and neoangiogenesis around endometriotic lesion. *U* Uterus, *US* uteroscacral ligaments. These can be seen to be edematous and inflamed. *E* endometriotic lesion, *BV* dilated blood vessels

reaction with ikntense angiogenesis around the lesion. Immune cells such as NK cells mediate the inflammatory reaction associated with endometriosis. It has been even been suggested that the peritoneal fluid may be an active promoter of growth of endometrial deposits by lipid peroxidation [9, 10]. These oxidants may stimulate endometrial-cell growth.

Endometriosis has many features of an autoimmune disease as there is increased polyclonal B-cell activity, high B-cell and T-cell counts, but with abnormal function [11, 12], and reduced natural-killer-cell activity [7, 12] High serum concentrations of IgG, IgA, and IgM autoantibodies [12, 13] and anti-endometrial antibodies have been reported [14]. Additionally, there is a high concordance of other autoimmune diseases or phenomena in women with endometriosis such as systemic lupus erythematosus, rheumatoid arthritis, Sjogren's syndrome, autoimmune thyroid disease, allergies, asthma and eczema [15]. Several studies have reported an association between autoimmune thyroid disease and endometriosis associated infertility, as shown by a high prevalence of positive anti-TPO antibodies [16].

Rather than being autoimmune in origin, the inflammatory reaction may be a response to the invasive properties of ectopic endomerium. Ectopic endometrial tissue fragments can attach to and invade the peritoneal surface. Matrix metalloproteinases (MMP) degrade extracellular matrix. MMP 7 and MMP 11 are normally expressed in the endometrium during menstrual breakdown and subsequent oestrogen-mediated endometrial proliferation. MMP 7 and MMP 11 are normally suppressed by progesterone during the secretory phase [17]. However, the ectopic endometrium is resistant to the effects of progesterone action due to non expression of the enzyme 17β hydroxysteroid dehydrogenase in women with endometriosis. Persistent expression of MMP might enable endometrial tissue to invade the peritoneal surface.

Survival of ectopic endometrial tissue in the peritoneal cavity is crucial for the establishment of viable implants. Downregulation of proapoptotic genes and upregulation of antiapoptotic genes of the *BCL2* and *BAX* families, have been reported endometriotic lesions [18]. Hence, there may be an intrinsic abnormality in endometriosis that permits ectopic endometrium to attach, survive, invade, and establish a blood supply.

Nuclear factor kappa B (NF-kB) may be crucial for mediating several biochemical processes associated with endometriosis [19]. NF-kB is activated by proinflammatory cytokines and oxidative stress and is increased in endometriotic lesions. NF-kB activation leads to the expression of a number of genes involved in inflammation, such as IL-1, IL-6, IL-8, and cyclooxygenase-2 [20]. Endometriotic tissue has been shown to activate NF-kB [21]. By activating proinflammatory genes, NF-kB perpetuates inflammation and macrophage recruitment. In addition to the inflammatory cascade, NF-kB regulates genes involved in antiapoptosis, tissue invasion, cell proliferation, and angiogenesis. In healthy women, NF-kB-DNA binding is decreased in the secretory phase relative to the proliferative phase, which may be due to the anti-inflammatory action of progesterone [21] . However, in women with endometriosis, NF-kB-DNA binding remains elevated during the secretory phase [22].

2.1 *Effect of Sex Hormones on Inflammatory Reaction*

Hormonal alterations may influence the ability of endometriotic cells to proliferate, attach to the mesothelium and evade immune mediated clearance [23]. Figure 9.2 shows the effect of various hormones and medications on the development of

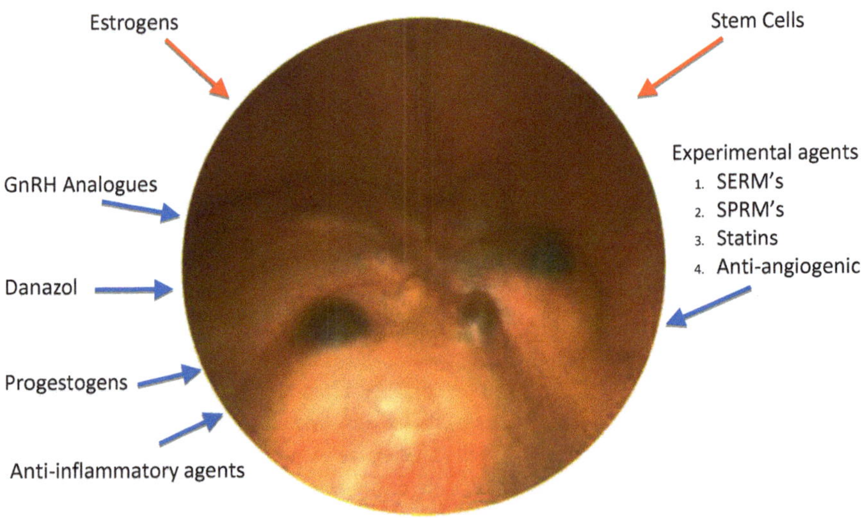

Fig. 9.2 Medications acting on endometriosis. *Red arrows* = stimulate endometriosis. *Blue arrows* indicate agents inhibiting endometriosis. *SERM's* = selective estrogen receptor modulators, *SPRMS* = selective progesterone receptor modulators

endometriosis. Endometriotic implants have increased expression of aromatase. and decreased expression of 17β-hydroxysteroid dehydrogenase [24] The consequence is a marked increase in the locally bioavailable estradiol concentration. Estradiol stimulates the production of prostaglandin E_2 which further stimulates aromatase activity [25]. In addition to estrogen dependence and biosynthesis, there is progesterone resistance in the pathophysiology of endometriosis [26]. Endometriotic lesions have reduced progesterone receptor expression when compared to eutopic endometrium, and an absence of progesterone receptor-B [27]. It has been reported that Akt (protein kinase) may be overactive, thus promoting progestogen resistance, and promotes the survival and proliferation of the diseased cells [28]. Increased Akt may be sufficient to downregulate progesterone receptor protein expression.

HOX genes, are dynamically expressed in the endometrium, where they are necessary for endometrial growth, differentiation, and implantation. In the human endometrium, the expression of HOXA10 and HOXA11 has peak expression at implantation in response to rising progesterone levels. However, maximal HOXA10 and HOXA11 expression does not occur in endometriosis, due to altered progesterone receptor expression or a dysregulated progesterone response. Consequently, other mediators of endometrial receptivity that are regulated by HOX genes, such as pinopodes, αvβ3 integrin, and IGFBP-1, are downregulated in endometriosis. HOXA10 hypermethylation may silence HOXA10 gene expression and account for decreased HOXA10 in the endometrium of women with endometriosis. Silencing of progesterone target genes by methylation is an epigenetic mechanism that mediates progesterone resistance. The relatively permanent nature of methylation may explain the widespread failure of treatments for endometriosis-related infertility [29].

Normal apoptosis mechanisms are suppressed in endometrial cells from women with endometriosis and within endometriotic lesions. Increased estrogen availability due to local estradiol synthesis and increased estrogen sensitivity lead to exaggerated protein kinase (Akt) activation and apoptosis inhibition [30].

2.2 *Innervation of Endometriosis*

In endometriosis there appears to be neoneurogenesis. The functional layer of eutopic endometrium (functionalis and basalis) is highly innervated in endometriosis [31] as are the ectopic lesions [32]. Small nerve fiber density assessed in endometrial biopsies from women with early stage endometriosis was 14 times higher than in biopsies from healthy controls [33]. Pain severity scores correlated with the presence of nerve fibers in both peritoneal and rectovaginal endometriotic lesions [34]. Persistent stimulation of nascent nociceptors in endometriotic lesions by inflammatory mediators may lead to central sensitization and neuropathic pain [35].

3 Evasion of Immune Clearance

Normally, endometrial cells which have been shed into the peritoneum are cleared by the immune system. Failure of the clearance mechanism may predispose to the implantation and growth of endometriosis. Endometrial cells have been found to be resistant to lysis by natural killer (NK) cells when compared to the endometrium from women without disease [7]. Shedding of intercellular adhesion molecule-1 (ICAM-1) by endometrial stromal cells from women with endometriosis may be one mechanism whereby these cells escape NK mediated clearance [36]. Impaired NK cell function (by downregulation of the NK1receptor and compromised macrophage function in endometriosis may further contribute to decreased clearance of lesions.

4 Treatment

No treatment of endomeriosis is curative, but based on tiding the patient over with pain relief, or optimising the possibility of fertility. The principles of treatment are summarised in Table 9.1 and include:- 1. Debulking and restoration of anatomy, by surgery. 2. Reduction of the estrogen required to maintain endometriosis. This can be achieved with GnRH analogs or Danazol 3. Reduction of the inflammatory response. This approach uses progestogens or anti-inflammatory agents. The object of medical management is to control pain and prevent recurrence of endometriosis. Medical treatment can be used on its own or as an adjunct to surgery.

4.1 Surgical Treatment

Surgical excision or ablation and adhesiolysis is an effective initial approach to treatment, whether for pain relief, or to enhance fertility [37]. There is widespread consensus that lesions should be excised where possible, especially deep endometriotic lesions [38]. Moreover, the surgery may be complex, particularly if other

Table 9.1 Management of endometriosis: therapeutic intervention

Debulking	Surgery
Estrogen reduction	1. Danazol 2. GnRH analogues 3. GnRH antagonists
Reducing inflammatory response	1. Progestogens 2. Oral contraceptive pills 3. Anti-inflammatory agents
Endometrial atrophy	1. LNG-IUS 2. Gestrinone

organs are involved such as bowel bladder or ureter. Moreover, even after excision which seems to be complete, there is often recurrence. Recurrence has been reported to vary from 10 to 55 % within 12 months [39], with recurrence affecting approximately 10 % of the remaining women each additional year [40]. The necessity for repeat surgery is higher in women younger than 30 years at the time of surgery [41]. Initial surgery produces better results than subsequent surgical procedures [42].

When there is deep infiltrating endometriosis, there is a dilemma. Incomplete resection may reduce symptomatic outcomes [43], but very extensive surgery increases the risk of ureteric and rectal injuries etc. [44]. Hence, it is clear is that surgical experience and expertise are required for endometriosis surgery, and that surgery should preferably be undertaken in centers of expertise.

Observational studies have suggested improved pain outcomes for women who undergo hysterectomy for Stage IV endometriosis [45]. However, hysterectomy is obviously reserved for women who do not require fertility. 96.9 % of women become asymptomatic after menopause [46]. It should also be borne in mind that as long as the patient has endogenous estrogen production, surgery is not curative, and in most cases only cytoreductive.

4.2 Progestogens

4.2.1 Mode of Action

Progestogens have been used to control the pain of endometriosis for many years. Progestogens are effective at a number of different levels. Some progestogens have anti-gonadotropic actions effect which inhibits ovarian function to create a hypoestrogenic environment. It was assumed that by directly acting on endometrial progesterone receptors progestogens may induce decidualization and atrophy of endometriosis. Progestogens can also reduce the inflammatory response. Hence, progestogens are frequently used as first-line therapy for the treatment of endometriosis. Progestogens do not reduce estradiol levels as much as GnRH agonists. Progestogens are advantageous over GnRH analogs as progestogens do not induce a medical menopause and are not associated with hot flushes and other hypoestrogenic side effects such as decreased bone mineral density.

The concept of progestogens acting on progesterone receptors and inducing decidualization as in normal endometrium does not seem to be valid. Endometriotic foci contain very few progesterone receptors [47]. The enzyme 17β hydroxysteroid dehydrogenase is absent in endometriotic tissue [48] and cannot be activated by progestogens. Progestogens also reduce the synthesis of their own receptors, which down regulates the sensitivity of the implants during long-term treatment.

Progestogens most probably acts by reducing the inflammatory response. TNF-α and estradiol induce the proliferation of endometriotic stroma cells via NF-kB, whereas progestogens reduce TNF-α induced NF-kB activation [49]. Progesterone itself is associated with decreased IFN-γ & increased IL-10 in endocervical fluid [50], upregulation of LIF mRNA expression in vitro. [51], and inhibits NK cell activity. The synthetic progestogen dydrogesterone has also been shown to modulate immune responses via suppression of IL-8 production in lymphocytes, inhibition of IFN-γ and increasing levels of IL-4 [52]. The increase in nitric oxide production seen with dydrogesterone may also play an important anti-inflammatory role [53].

4.2.2 Different Progestogens

Many progestogens have been have been used for the treatment of endometriosis. These include medroxyprogesterone acetate (administered by intramuscular injection), desogestrel, dienogest, cyproterone acetate, dydrogesterone etc. administered orally, either alone, or in combination with oestrogens in the combined oral contraceptive pill or levonorgestrel absorbed from an intrauterine contraceptive device. Progestogens have many beneficial effects. However, the results do not differ very much from the use of the combined estrogen progestin oral contraceptive pill. Some specific progestogens are described below.

Dydrogesterone

Dydrogesterone is a stereoisomer of progesterone manufactured by treating progesterone with ultraviolet light. Dydrogesterone stimulates the progesterone receptor directly without affecting progesterone levels. Dydrogesterone also binds the receptor 50 % more than progesterone itself [54]. Dydrogesterone does not stimulate the androgen, glucocorticoid or estrogen receptor.

There are numerous small comparative studies and case reports of dydrogesterone in endometriosis since the 1960s. Schweppe [55] summarized seven control studies of dydrogesterone. Doses between 10 and 60 mg/day, were used for various numbers of days per cycle, and over periods of 3–9 months. The majority of women became symptom-free or experienced a significant reduction in the number/severity of symptoms. These findings have been supported by laparoscopic examination in several of the studies. There was an overall success rate of approximately 70–90 %. Laparoscopic examination in several of the studies supported these findings. However, a Cochrane review published in 2012 [56] which included 13 randomised controlled trials evaluating the use of progestogens, included found only one RCT for dydrogesterone [57]. Sixty-two women were randomized to dydrogesterone or placebo. Only 39 women completed the study and underwent a second look laparoscopy. There was no significant improvement in objective efficacy at 6 months compared to placebo (OR 0.53, 95 % CI 0.14–1.94) Nor were there any differences observed in the change in pain score at 12 months of follow up (OR 0.80, 95 % CI 0.27–2.37; NS). However, the wide confidence intervals and small number of patients indicates that Schweppe's [55] figures are probably more relevant.

Cyclic application of dydrogesterone has also been shown to induce regular menstruation with reduced blood loss and fewer days of bleeding, combined with excellent symptomatic relief, in women suffering from dysmenorrhea. Dydrogesterone is especially useful in patients desiring pregnancy Because dydrogesterone does not inhibit ovulation it can be used for symptomatic treatment of pain, reduction of bleeding problems and cycle control.

Dienogest

Dienogest is a synthetic orally active progestogen. As dydrogesterone, dienogest is highly selective for the progesterone receptor and exerts a strong progestational effect. However, dienogest differs from dydrogesterone in that it has a moderate antagonist action on the androgen receptor, and has a moderate antigonadotropic effect [58]. However, there is no androgenic, glucocorticoid, mineralocorticoid activity or estrogenic activity. The therapeutic dose (2 mg) inhibits ovulation in healthy women with normal menstrual cycles [59]. However, dienogest only moderately reduces oestrogen levels, hence, dienogest does affect bone mineral density [60]. Dienogest does not reduce sex hormone-binding globulin, is bound unspecifically to albumin and does not accumulate using oral doses of 2 mg/day [61].

Two prospective placebo controlled randomized studies assessed dienogest 2 mg daily against placebo [58] or versus leuprorelin depot [60]. Both trials showed a significant improvement in endometriosis-related symptoms, and a similar effectiveness to GnRH agonist therapy. However, the bleeding pattern differed substantially between the two groups. In the leuprolide group most women had infrequent bleeding in the first 90 days and amenorrhea after prolonged treatment. In the dienogest group prolonged and irregular bleeding were frequent in the first 90 days of treatment. Bleeding problems occurred in up to 80 % of patients within the first 3 months of treatment.

In a single-arm extension study of treatment for 15 months and follow-up 6 months after discontinuation of treatment, dienogest was shown to reduce pain symptoms with normalisation and long term relief of symptoms even after treatment discontinuation.

Dienogest has a good safety and efficacy profile, with good tolerability, antiandrogenic action and weak antigonadotropic activity, combined with typical characteristics of 19-norprogestins: strong suppressive action on the endometrium in low doses, a short half-life and high bioavailability [62]. However, there are no trials comparing the efficacy and tolerability of dienogest to OCPs or other progestogens.

Medroxy Progesterone Acetate (MPA)

MPA can be administered orally in a dose of 15–50 mg orally or injected as a depot form (DMPA). Bergqvist et al. [63] compared MPA to placebo. There was a greater quality of life after MPA and pain relief. Telimaa et al. [64] reported the results of a prospective, randomized trial comparing MPA to Danazol. A 50 % regression rate of ectopic implants and 13 % partial regression with scar formation was reported in the treatment group compared to 12 and 6 %, respectively, in the placebo group, and a net reduction in pain symptoms after treatment compared to placebo. (OR=0.70, CI −8.61 to −5.39; P<0.00001). When Danazol and MPA were compared, both alleviated endometriosis-associated pelvic pain, lower back pain and defecation pain, but they did not differ from each other in these actions. The authors concluded that because of good efficacy and tolerance, high-dose MPA is a useful alternative in the hormonal treatment of endometriosis. However, MPA has glucocorticoid and androgenic effects. Brown et al. [56], have reported significantly more cases of acne (six versus one) and oedema (11 versus one) in the medroxyprogesterone acetate group compared with placebo. The dose is 20 mg to 100 mg daily. Harrison and Barry-Kinsella [65] published the results of a placebo controlled trial. Initial and second-look laparoscopy were performed to grade the lesions according to the revised American Fertility Society stages. Surprisingly, both MPA and placebo therapy achieved similar statistically significant reductions in stages and scores at second-look laparoscopy. However, MPA was more effective in improving overall well-being. The authors concluded, that as both MPA and placebo were equally effective in treating endometriosis over a 3-month period, and questioned the role of using MPA altogether.

MPA can also be administered intramuscularly in a depot form (DPMA). DPMA is long acting, and a 150 mg. dose may only need to be repeated after 3 months. Vercellini [66] compared 150 mg of depot medroxyprogesterone (DMPA) every 3 months with a 20 μg oral contraceptive pill (OCP) with 50 mg danazol. Both the pill and danazol were taken for 3 weeks out of 4. The primary endpoint was the degree of satisfaction at the end of therapy. Pain reduction with MPA was as effective as danazol. DPMA has also been compared to leuprolide [67]. Symptoms of dysmenorrhoea were significantly reduced in the DMPA group at 6 months compared with leuprolide (OR 0.19, 95 % CI 0.05–0.69; P=0.01) but this effect was short lived, and not present at the 12 months follow-up (OR 0.63, 95 % CI 0.37–1.08). There were no differences between groups at 12 months follow up, regarding dyspareunia after 6 months. At 12 months fewer women in the leuprolide group reported dyspareunia (OR 4.83, 95 % CI 2.14–10.93).

Side effects include breakthrough bleeding in approximately 40 % of patients, nausea, breast tenderness, fluid retention and depression.

Cyproterone Acetate

Vercellini [68] compared 12.5 mg cytoproterone acetate daily to a continuous monophasic OCP once daily (0.02 μg ethinyl estradiol and 0.15 mg desogestrel). The primary endpoint, as in their previous study, was the degree of satisfaction at the end of therapy. A change in severity of symptoms was also measured by a 100 mm visual analogue score and a 0–3 point verbal rating scale. Cyproterone however, has significant anti-androgenic effects.

Levonorgestrel Intrauterine System (LNG-IUS)

The LNG-IUS is a contraceptive intrauterine device (IUD). As it releases norgesterel in a constant fashion, it lessens the excess bleeding associated with other IUD's, and may even lead to amenorrhea. The LNG-IUS is therefore indicated for women with menorrhagia and dysfunctional bleeding. The constant release of levonorgestrel leads to mean plasma concentrations between 100 and 200 pg/ml [69]. Endometrial exposure to LNG induces endometrial atrophy. Hence the LNG-IUS can only be used for endometriosis in women who do not desire fertility, and who are prepared to accept amenorrhea.

Several small RCTs have compared the use of LNG-IUS in endometriosis to GnRH agonists and Depot medroxyprogesterone acetate [70]. The mechanism by which the LNG-IUS decreases endometriosis related symptoms is unclear, as the LNG-IUS does not inhibit ovulation nor does it induce a hypoestrogenic state. It has been suggested that the LNG-IUS acts by decreasing the expression of glandular and stromal estrogen and progesterone receptors in the ectopic endometrium [71]. With the LNG-IUS a reduction of the severity of endometriosis has been seen at laparoscopy [72, 73]. The echographic size of recto-vaginal lesions under has also been seen on ultrasound under LNG-IUS treatment [74]. The LNG-IUS has been shown to reduce pain [75]. Tanmahasamut [76] compared treatment with the LNS-IUS after laparoscopic conservative surgery to expectant management. There was a significant reduction in dysmenorrhoea [5.0 cm vs 8.1 cm on the visual analogue scale (VAS)] and non-cyclic pelvic pain (VAS 2.2 cm vs 4.8 cm) but no effect on dyspareunia. The LNG-IUS is as effective as DMPA with no impact on bone mineral density [77]. Bayoglu et al. [78] compared the efficacy of the LNG-IUS with the GnRH analogue gosareline on endometriosis related chronic pelvic pain in patients with severe endometriosis during 12 months. Both treatment modalities showed comparable effectiveness in the treatment of chronic pelvic pain related endometriosis.

Norethindrone Acetate (NETA)

Norethindrone (Norethisterone) acetate can be used continuously in a dose of 2.5 mg per day. The evidence for pain relief comes from a study by Muneyyirci-Delale and Karacan [79]. Fifty-two women with endometriosis confirmed by laparoscopy were treated with NETA. Dysmenorrhea and noncyclic pelvic pain were relieved in 48/52 (92.3 %) and 25/28 (89.2 %) of patients, respectively. Similar results were found in recurrent endometrioma [80], and adenomyosis. The advantages of NETA include, excellent cycle control, and no harmful effect on the lipoprotein profile [81]. Ferrero et al. [82] has shown that NETA can relieve the pain of endometriosis, particularly when that pain presents as dysmenorrhea.

Effect of Progestogens on Endometriosis Related Infertility

Numerous mechanisms have been proposed to explain the effect of rndometriosis in infertility:- altered folliculogenesis, reduced preovulatory steroidogenesis by granulosa cells, decreased capability etc. decreased capability of fimbrial ovum capture, sperm phagocytosis by peritoneal and oviductal macrophages, anti-sperm antibodies and reduced sperm penetration and velocity In addition, altered egg–sperm interaction, defective implantation and impaired early embryonic development have been reported to explain endometriosis related infertility. Consequently it is not surprising that although medical management improves the quality of life for many women with endometriosis, the effect on endometriosis related fertility is not so successful. Many progestational agents inhibit ovulation, precluding their use in patients desiring fertility. A Cochrane database metaanalysis [83] which included 23 trials of over 3,000 women found that pretreatment with ovulation inhibiting agents such as oral contraceptives, progestogens, danazol etc. does not improve long term fecundity, and only delays conception. Pregnancy rates following progestin therapy however, depend on the stage of the disease, and whether medical therapy is an adjunct to surgery. To date, there is no randomized controlled trial that has shown an improvement in fertility after any progestin medication. The situation may be different with progestogens which do not affect ovulation such as dydrogesterone. As stated above, dydrogesterone has been shown to improve symptoms and lead to improvement of the findings at laparoscopy. However, in women with infertility and severe disease, there is little evidence of effect on fecundity. For fertility enhancement, surgical treatment is probably preferable, with assisted reproduction immediately after surgery, prior to the recurrence of disease.

4.3 Oral Contraceptive Pill (OCP)

The estrogenic component of the OCP prevents ovulation while the progestogen is continuously supplied. A therapeutic trial is easily performed with either continuous or cyclic OCPs. Continuous administration of OCPs, avoids menstruation and its associated pain. However, absence of menstruation is not always acceptable to all women. Like progestogen regimens, the progestogen in the pill is believed to produce initial decidualization followed by atrophy. To-day the OCP is the most commonly prescribed treatment for endometriosis symptoms. A study by Guzick et al. [84] compared Lupron and continuous oral contraceptives for the treatment of endometriotic pain; both were found to be equally effective. An advantage of OCP's is that women with endometriosis are at increased risk of epithelial ovarian carcinoma. OCPs may prevent this tumor [85].

4.4 Anti-progestogens (Gestrinone)

Gestrinone is a synthetic steroid with mixed progestogen and antiprogestogen effects, some mild androgenic activity, and some anti-estrogenic activity. The mechanism of action consists of suppression of the release of pituitary gonadotropins. Gestrinone also interacts with the endometrium, inhibiting its growth. Gestrinone enhances lysosomal degradation, leading to a rapid decrease in progesterone receptors. Gestrinone is administered orally in a dose of 2.5–10 mg. weekly. A literature search in the Cochrane database [56] found no RCT comparing gestrinone to placebo or no treatment. The Gestrinone Italian Study Group [86] showed that oral gestrinone was as effective as leuprolide depot injection for pain relief. Visual analog scale pain scores decreased from 4.07 ± 2.86 to 1.23 ± 2.65 at 6 months in the gestrinone group. Two studies have compared gestrinone to danazol [87, 88]. There was no difference in either group regarding pain. Side effects are due to the androgenic and anti-estrogenic effects including, voice changes, hirsutism, and clittoral enlargement. However, most side effects are mild, transient and reversible.

4.5 GnRH Analogs

GnRH agonists are decapeptides that differ from endogenous GnRH by the substitution of one or more amino acids. GnRH agonists exert a continuous potent action on GnRH receptors leading to deep suppression of gonadotrophin secretion, ovarian suppression, hypoestrogenism, and a pseudo menopausal state. The pseudomenopausal state leads to side effects such as hot flushes, headaches, vaginal dryness etc. and more serious effects such as bone loss. The lack of estrogen leads to remission of endometriosis during treatment. Due to frequent adverse effects, GNRH agonists are usually used as second-line treatment for endometriosis when the condition is refractory to other forms of therapy such as OCP or progestogens. Therefore, if GnRH analogs are used, concomitant add back therapy with estrogen and progestogens is indicated. The addition of estrogen and progestogen to the GnRH agonist does not seem to reduce the efficacy of GnRH agonists on pain symptoms [89]. The efficacy of GnRH agonists on endometriotic pain has been confirmed in a Cochrane database systematic review [90]. The metaanalysis included 41 trials comprising 4,935 women. The results of the metaanalysis showed that GnRH agonists are more efficient than no treatment or placebo for pain relief. However, GnRH agonists have not been shown to be superior to danazol or levonorgestrel.

GnRH antagonists do not initially release FSH and LH as the agonists, but the antagonist competes with GnRH for the receptor inhibiting any gonadotrophin release. However, the results of high quality trials are awaited/.

4.6 Danazol

Danazol is a synthetic androgen that acts by inducing atrophy of endometriotic implants. Danazol was the drug of choice for treating endometriotic pain in the 1970s and 1980s. Danazol is used in a dose of 600–800 mg per day. A Cochrane database review analysed five studies on danazol and concluded that danazol is effective for endometriotic pain compared to placebo or no treatment and improves the extent of endometriosis on laparoscopy [91]. Danazol declined in popularity due to the side effects such as weight gain, lipid profile alterations, acne, hirsutism, alopecia, mood changes, deepening of the voice, etc.

4.7 Choice of Treatment

Progestogens seem to be effective in treating endometriosis-related pain. There is however no evidence for an improved efficacy compared to other hormonal treatment and their use in the treatment of symptomatic endometriosis should be conditional to patient acceptability and tolerance of side effects. There is a problem in that the different progestogens have never been compared to one another. The LNG-IUS has been compared to Depot medroxyprogesterone acetate [70], gestrinone to leuprolide [86], and danazol [87, 88]. Danazol has been compared to medroxyprogesterone acetate [64], but there are few other comparative trials. Therefore the choice of progestogen dependant on the patients desire for conception, and often the treating physician's personal experience and preference.

The World Endometriosis Society published a Consensus on the management of endometriosis. Their recommendations are outlined in Table 9.2. However, there is no panacea for the treatment of endometriosis. The choice of treatment is dependent

Table 9.2 Principles of management of endometriosis (adapted from [92])

Treatment	Principles for usage
Surgical treatment	1. Usually first line approach 2. Definitive diagnosis can be made 3. Depends on acceptability to patient 4. Improves fertility
Medical treatment	1. Provides symptomatic relief 2. Used for treatment of recurrences 3. Used while awaiting surgery 4. May be treatment of choice in adolescents or after menopause 5. No evidence of improvement of infertility
First line medical treatment	1. NSAIDS, OCP, progestogens 2. Choice of progestogen depends on whether ovulation should be inhibited or desire for concurrent fertility
Second line medical treatment	GnRH agonists with add back therapy, LNG-IUS

on the patients age, whether the main symptom is pain or difficulty conceiving, acceptability of the side effects of medical treatment, acceptability of surgery and the results of previous therapy. At present a number of new agents are being assessed for treating endometriosis (Table 9.2). These include selective estrogen receptor modulators, selective progesterone receptor modulators, statins and anti-angiogenic drugs. However, it is too soon to judge their efficacy or give recommendations.

References

1. Cornillie FJ, Oosterlynck D, Lauweryns JM, et al. Deeply infiltrating pelvic endometriosis: histology and clinical significance. Fertil Steril. 1990;53(6):978–83.
2. Eskenazi B, Warner ML. Epidemiology of endometriosis. Obstet Gynecol Clin North Am. 1997;24:235–58.
3. Meuleman C, Tomassetti C, D'Hoore A, Van Cleynenbreugel B, Penninckx F, Vergote I, D'Hooghe T. Surgical treatment of deeply infiltrating endometriosis with colorectal involvement. Hum Reprod Update. 2011;17:311–26.
4. Leibovic DI, Mueller MD, Taylor RN. Immunobiology of endometriosis. Fertil Steril. 2001;75:1–10.
5. Sharpe-Timms KL, Piva M, Ricke EA, Surewicz K, Zhang YL, Zimmer RL. Endometriosis synthesizes and secretes a haptoglobin-like protein. Biol Reprod. 1998;58:988–94.
6. Piva M, Horowitz GM, Sharpe-Timms KL. Interleukin-6 differentially stimulates haptoglobin production by peritoneal and endometriotic cells in vitro: a model for endometrium-peritoneum interaction in endometriosis. J Clin Endocrinol Metab. 2001;86:2553–61.
7. Oosterlynck DJ, Cornillie FJ, Waer M, Vandeputte M, Koninckx PR. Women with endometriosis show a defect in natural killer activity resulting in a decreased cytotoxicity to autologous endometrium. Fertil Steril. 1991;56:45–51.
8. Taylor RN, Lebovic DI, Meuller MD. Angiogenic factors in endometriosis. Ann N Y Acad Sci. 2001;955:89–100.
9. Santanam N, Murphy AA, Parthasarathy S. Macrophages, oxidation, and endometriosis. Ann N Y Acad Sci. 2001;955:183–200.
10. Van Langendonckt A, Casanas-Roux F, Donnez J. Oxidative stress and peritoneal endometriosis. Fertil Steril. 2002;77:861–70.
11. Eisenberg VH, Zolti M, Soriano D. Is there an association between autoimmunity and endometriosis? Autoimmun Rev. 2012;11(11):806–14.
12. Nothnick WB. Treating endometriosis as an autoimmune disease. Fertil Steril. 2001;76:223–40.
13. Gleicher N, El-Roeiy A, Confino E, Friberg J. Is endometriosis an autoimmune disease? Obstet Gynecol. 1987;70:115–22.
14. Grossinkinsky CM, Halme J. Endometriosis: the host response. Baillieres Clin Obstet Gynaecol. 1993;7:701–13.
15. Sinaii N, Cleary SD, Ballweg ML, Nieman LK, Stratton P. High rates of autoimmune and endocrine disorders, fibromyalgia, chronic fatigue syndrome and atopic diseases among women with endometriosis: a survey analysis. Hum Reprod. 2002;17:2715–24.
16. Poppe K, Glinoer D, Van Steirteghem A, Tournaye H, Devroey P, Schiettecatte J, et al. Thyroid dysfunction and autoimmunity in infertile women. Thyroid. 2002;12:997–1001.
17. Osteen KG, Keller NR, Feltus FA, Melner MH. Paracrine regulation of matrix metalloproteinase expression in the normal endometrium. Gynecol Obstet Invest. 1999;48 suppl 1:2–13.
18. Garcia-Velasco JA, Arici A. Apoptosis and the pathogenesis of endometriosis. Semin Reprod Med. 2003;21:165–72.

19. Guo SW. Nuclear factor-kappab (NF-kappaB): an unsuspected major culprit in the pathogenesis of endometriosis that is still at large? Gynecol Obstet Invest. 2007;63:71–97.
20. Perkins ND. Integrating cell-signalling pathways with NF-kappaB and IKK function. Nat Rev Mol Cell Biol. 2007;8:49–62.
21. Gonzalez-Ramos R, Donnez J, Defrere S, et al. Nuclear factor-kappa B is constitutively activated in peritoneal endometriosis. Mol Hum Reprod. 2007;13:503–9.
22. Gonzalez-Ramos R, Rocco J, Rojas C, et al. Physiologic activation of nuclear factor kappa-B in the endometrium during the menstrual cycle is altered in endometriosis patients. Fertil Steril. 2012;97:645–51.
23. Kitawaki J, Kado N, Ishihara H, Koshiba H, Kitaoka Y, Honjo H. Endometriosis: the pathophysiology as an estrogen-dependent disease. J Steroid Biochem Mol Biol. 2002;83:149–55.
24. Zeitoun K, Takayama K, Sasano H, Suzuki T, Moghrabi N, Andersson S, et al. Deficient 17beta-hydroxysteroid dehydrogenase type 2 expression in endometriosis: failure to metabolize 17beta-estradiol. J Clin Endocrinol Metab. 1998;83:4474–80.
25. Noble LS, Takayama K, Zeitoun KM, Putman JM, Johns DA, Hinshelwood MM, et al. Prostaglandin E2 stimulates aromatase expression in endometriosis-derived stromal cells. J Clin Endocrinol Metab. 1997;82:600–6.
26. Bulun SE, Cheng YH, Yin P, Imir G, Utsunomiya H, Attar E, et al. Progesterone resistance in endometriosis: link to failure to metabolize estradiol. Mol Cell Endocrinol. 2006;248: 94–103.
27. Attia GR, Zeitoun K, Edwards D, Johns A, Carr BR, Bulun SE. Progesterone receptor isoform A but not B is expressed in endometriosis. J Clin Endocrinol Metab. 2000;85:2897–902.
28. Lee II, Kim JJ. Influence of AKT on progesterone action in endometrial diseases. Biol Reprod. 2014;91:63. pii: biolreprod.114.119255. [Epub ahead of print].
29. Cakmak H, Taylor HS. Implantation failure: molecular mechanisms and clinical treatment. Hum Reprod Update. 2011;17(2):242–53.
30. Reis FM, Petraglia F, Taylor RN. Endometriosis: hormone regulation and clinical consequences of chemotaxis and apoptosis. Hum Reprod Update. 2013;4:406–18.
31. Tokushige N, Markham R, Russell P, Fraser IS. High density of small nerve fibres in the functional layer of the endometrium in women with endometriosis. Hum Reprod. 2006;21:782–7.
32. Tokushige N, Markham R, Russell P, Fraser IS. Nerve fibres in peritoneal endometriosis. Hum Reprod. 2006;21:3001–7.
33. Bokor A, Kyama CM, Vercruysse L, Fassbender A, Gevaert O, Vodolazkaia A, De Moor B, Fülöp V, D'Hooghe T. Density of small diameter sensory nerve fibres in endometrium: a semi–invasive diagnostic test for minimal to mild endometriosis. Hum Reprod. 2009;24:3025–32.
34. McKinnon B, Bersinger NA, Wotzkow C, Mueller MD. Endometriosis-associated nerve fibers, peritoneal fluid cytokine concentrations, and pain in endometriotic lesions from different locations. Fertil Steril. 2012;97:373–80.
35. Stratton P, Berkley KJ. Chronic pelvic pain and endometriosis: translational evidence of the relationship and implications. Hum Reprod Update. 2011;17:327–46.
36. Somigliana E, Vigano P, Gaffuri B, Guarneri D, Busacca M, Vignali M. Human endometrial stromal cells as a source of soluble intercellular adhesion molecule (ICAM)-1 molecules. Hum Reprod. 1996;11:1190–4.
37. Jacobson TZ, Duffy JM, Barlow D, Koninckx PR, Garry R. Laparoscopic surgery for pelvic pain associated with endometriosis. Cochrane Database Syst Rev. 2009;4:CD001300.
38. Johnson NP, Hummelshoj L. Consensus on current management of endometriosis. Hum Reprod. 2013;28(6):1552–68.
39. Vercellini P, Somigliana E, Viganò P, De Matteis S, Barbara G, Fedele L. The effect of second-line surgery on reproductive performance of women with recurrent endometriosis: a systematic review. Acta Obstet Gynecol Scand. 2009;88:1074–82.
40. Guo SW. Recurrence of endometriosis and its control. Hum Reprod Update. 2009;15: 441–61.
41. Shakiba K, Bena JF, McGill KM, Minger J, Falcone T. Surgical treatment of endometriosis: a 7-year follow-up on the requirement for further surgery. Obstet Gynecol. 2008;111:1285–92.

42. Abbott J, Hawe J, Hunter D, Holmes M, Finn P, Garry R. Laparoscopic excision of endometriosis: a randomized, placebo-controlled trial. Fertil Steril. 2004;82:878–84.
43. Vercellini P, Pietropaolo G, De Giorgi O, Daguati R, Pasin R, Crosignani PG. Reproductive performance in infertile women with rectovaginal endometriosis: is surgery worthwhile? Am J Obstet Gynecol. 2006;195:1303–10.
44. Koninckx PR, Timmermans B, Meuleman C, Penninckx F. Complications of CO2-laser endoscopic excision of deep endometriosis. Hum Reprod. 1996;11:2263–8.
45. Ford J, English J, Miles WA, Giannopolous T. Pain, quality of life and complications following the radical resection of rectovaginal endometriosis. Br J Obstet Gynaecol. 2004;111:353–6.
46. Fagervold B, Jenssen M, Hummelshoj L, Moen MH. Life after a diagnosis with endometriosis—a 15 years follow-up study. Acta Obstet Gynecol Scand. 2009;88:914–9.
47. Kauppila A, Rönnberg L, Viehko R. Steroidrezeptoren in endometrischem Gewebe. Endometriose. 1986;4:56–9.
48. Vierikko P, Kauppila A, Rönnberg L, Viehko R. Steroidal regulation of endometriosis tissue. Fertil Steril. 1985;43:218–23.
49. Horie S, Harada T, Mitsunari M, Taniguchi F, Iwabe T, Terakawa N. Progesterone and progestational compounds attenuate tumor necrosis factor alpha-induced interleukin-8 production via nuclear kappa B inactivation in endometriotic stromal cells. Fertil Steril. 2005;83:1530–5.
50. Alimohamadi S, Javadian P, Gharedaghi MH, Javadian N, Alinia H, Khazardoust S, Borna S, Hantoushzadeh S. Progesterone and threatened abortion: a randomized clinical trial on endocervical cytokine concentrations. J Reprod Immunol. 2013;98(1–2):52–60.
51. Aisemberg J, Vercelli CA, Bariani MV, Billi SC, Wolfson ML, Franchi AM. Progesterone is essential for protecting against LPS-induced pregnancy loss. LIF as a potential mediator of the anti-inflammatory effect of progesterone. PLoS One. 2013;8(2):e56161.
52. Ragupathy R, Al Mutawa E, Makhseed M, Azizieh F, Szekeres-Bartho J. Modulation of cytokine production by dydrogesterone in lymphocytes from women with recurrent miscarriage. BJOG. 2005;112:1096–101.
53. Simoncini T, Caruso A, Giretti MS, Scorticati C, Fu XD, Garibaldi S, et al. Effects of dydrogesterone and of its stable metabolite, 20-alpha-dihydrodydrogesterone, on nitric oxide synthesis inhumanendothelial cells. Fertil Steril. 2006;86 Suppl 4:1235–42.
54. King RJ, Whitehead MI. Assessment of the potency of orally administered progestins in women. Fertil Steril. 1986;46(6):1062–6.
55. Schweppe KW. The place of dydrogesterone in the treatment of endometriosis and adenomyosis. Maturitas. 2009;65S:S23–7.
56. Brown J, Kives S, Akhtar M. Progestagens and anti-progestagens for pain associated with endometriosis. Cochrane Database Syst Rev. 2012;3:CD002122. doi:10.1002/14651858. CD002122.pub2.
57. Overton CE, Lindsay PC, Johal B, Collins SA, Siddle NC, Shaw RW, Barlow DH. A randomized, double-blind, placebo-controlled study of luteal phase dydrogesterone (Duphaston) in women with minimal to mild endometriosis. Fertil Steril. 1994;62(4):701–7.
58. Strowitzki T, Faustmann T, Gerlinger C, et al. Dienogest in the treatment of endometriosis-associated pelvic pain: a 12-week, randomized, double-blind, placebo-controlled study. Eur J Obstet Gynecol Reprod Biol. 2010;151:193–8.
59. Klipping C, Duijkers I, Remmers A, et al. Ovulation-inhibiting effects of dienogest in a randomized, dose-controlled pharmacodynamic trial of healthy women. J Clin Pharmacol. 2011;52:1704–13.
60. Strowitzki T, Marr J, Gerlinger C, et al. Dienogest is as effective as leuprolide acetate in treating the painful symptoms of endometriosis: a 24-week, randomized, multicentre, open-label trial. Hum Reprod. 2010;25:633–41.
61. Kuhl H. Comparative pharmacology of newer progestogens. Drugs. 1996;51:188–215.
62. Zimmermann H, Duvauchelle T, Gualano V, Kaufmann G, Bervoas-Martin S, Breitbarth H. Pharmacokinetics of dienogest as a single drug or in combination with estradiol valerate or ethinylestradiol. Drugs Today. 1999;35(Suppl C):27–39.

63. Bergqvist A, Theorell T. Changes in quality of life after hormonal treatment of endometriosis. Acta Obstet Gynecol Scand. 2001;80:628–37.
64. Telimaa S, Puolakka J, Ronnberg L, et al. Placebo-controlled comparison of danazol and high-dose medroxyprogesterone acetate in the treatment of endometriosis. Gynecol Endocrinol. 1987;1:13–23.
65. Harrison RF, Barry-Kinsella C. Efficacy of medroxyprogesterone treatment in infertile women with endometriosis: a prospective, randomized, placebo-controlled study. Fertil Steril. 2000;74:24–30.
66. Vercellini P, De Giorgi O, Oldani S, Cortesi I, Panazza S, Crosignani PG. Depot medroxyprogesterone acetate versus an oral contraceptive combined with very-low-dose danazol for long-term treatment of pelvic pain associated with endometriosis. Am J Obstet Gynecol. 1996;175:396–401. Brown.
67. Sclaff W, Carson S, Luciano A, Ross D, Bergvist A. Subcutaneous injection of depot medoxyprogesterone acetate compared with leuprolide acetate in the treatment of endometriosis associated pain. Fertil Steril. 2006;85(2):314–25.
68. Vercellini P, De Giorgi O, Mosconi P, Stellato G, Vicentini S, Crosignani PG. Cyproterone acetate versus a continuous monophasic oral contraceptive in the treatment of recurrent pelvic pain after conservative surgery for symptomatic endometriosis. Fertil Steril. 2002;77(1):52–61.
69. Luukkainen T, Lahteenmaki P, Toivonen J. Levonorgestrel-releasing intrauterine device. Ann Med. 1990;22:85–90.
70. Streuli I, de Ziegler D, Santulli P, Marcellin L, Borghese B, Batteux F, Chapron C. An update on the pharmacological management of endometriosis. Expert Opin Pharmacother. 2013;14(3):291–305.
71. Engemise SL, Willets JM, Taylor AH, Emembolu JO, Konje JC. Changes in glandular and stromal estrogen and progesterone receptor isoform expression in eutopic and ectopic endometrium following treatment with the levonorgestrel-releasing intrauterine system. Eur J Obstet Gynecol Reprod Biol. 2011;157:101–6.
72. Lockhat FB, Emembolu JO, Konje JC. The evaluation of the effectiveness of an intrauterine administered progestogen (levonorgestrel) in the symptomatic treatment of endometriosis and in the staging of the disease. Hum Reprod. 2004;19:179–84.
73. Vercellini P, Frontino G, De Giorgi O, et al. Comparison of a levonorgestrel-releasing intrauterine device versus expectant management after conservative surgery for symptomatic endometriosis: a pilot study. Fertil Steril. 2003;80:305–9.
74. Fedele L, Bianchi S, Zanconato G, et al. Use of a levonorgestrel-releasing intrauterine device in the treatment of rectovaginal endometriosis. Fertil Steril. 2001;75:485–8.
75. Serachioli R, Mabrouk M, Frascà C, Manuzzi L, Savelli L, Venturoli S. Long-term oral contraceptive pills and postoperative pain management after laparoscopic excision of ovarian endometrioma: a randomized controlled trial. Fertil Steril. 2010;94:464–71.
76. Tanmahasamut P, Rattanachaiyanont M, Angsuwathana S, et al. Postoperative levonorgestrel-releasing intrauterine system for pelvic endometriosis-related pain: a randomized controlled trial. Obstet Gynecol. 2012;119:519–26.
77. Wong AY, Tang LC, Chin RK. Levonorgestrel-releasing intrauterine system (Mirena) and Depot medroxyprogesterone acetate (Depoprovera) as long-term maintenance therapy for patients with moderate and severe endometriosis: a randomised controlled trial. Aust N Z J Obstet Gynaecol. 2010;50:273–9.
78. Bayoglu Tekin Y, Dilbaz B, Altinbas SK, et al. Postoperative medical treatment of chronic pelvic pain related to severe endometriosis: levonorgestrel-releasing intrauterine system versus gonadotropin-releasing hormone analogue. Fertil Steril. 2011;95:492–6.
79. Muneyyirci-Delale O, Karacan M. Effect of norethindrone acetate in the treatment of symptomatic endometriosis. Int J Fertil Womens Med. 1998;43:24–7.
80. Muneyyirci-Delale O, Anopa J, Charles C, Mathur D, Parris R, Cutler JB, Salame G, Abulafia O. Medical management of recurrent endometrioma with long-term norethindrone acetate. Int J Womens Health. 2012;4:149–54.

81. Riis BJ, Lehmann HJ, Christiansen C. Norethisterone acetate in combination with estrogen: effects on the skeleton and other organs. A review. Am J Obstet Gynecol. 2002;187:1101–16.
82. Ferrero S, Camerini G, Ragni N, Venturini PL, Biscaldi E, Seracchioli R, Remorgida V. Letrozole and norethisterone acetate in colorectal endometriosis. Eur J Obstet Gynecol Reprod Biol. 2010;150:199–202.
83. Hughes E, Brown J, Collins JJ, Farquhar C, Fedorkow DM, Vandekerckhove P. Ovulation suppression for endometriosis. Cochrane Database Syst Rev. 2007;3:CD000155. Review.
84. Guzick DS, Huang LS, Broadman BA, Nealon M, Hornstein MD. Randomized trial of leuprolide versus continuous oral contraceptives in the treatment of endometriosis-associated pelvic pain. Fertil Steril. 2011;95(5):1568–73.
85. Modugno F, Ness RB, Allen GO, Schildkraut JM, Davis FG, Goodman MT. Oral contraceptive use, reproductive history, and risk of epithelial ovarian cancer in women with and without endometriosis. Am J Obstet Gynecol. 2004;191(3):733–40.
86. Gestrinone Italian Study Group. Gestrinone versus a gonadotropin-releasing hormone agonist for the treatment of pelvic pain associated with endometriosis: a multicenter, randomized, double-blind study. Fertil Steril. 1996;66:911–9.
87. Bromham DR, Booker MW, Rose GL, Wardle PG, Newton JR. A multicentre comparative study of gestrinone and danazol in the treatment of endometriosis. J Obstet Gynaecol. 1995;15:188–94.
88. Fedele L, Arcaini L, Bianchi S, Viezzoli T, Arcaini L, Candiani GB. Gestrinone versus danazol in the treatment of endometriosis. Fertil Steril. 1989;51(5):781–5.
89. Surrey ES. Gonadotropin-releasing hormone agonist and add-back therapy: what do the data show? Curr Opin Obstet Gynecol. 2010;22:283–8.
90. Brown J, Pan A, Hart RJ. Gonadotrophin-releasing hormone analogues for pain associated with endometriosis. Cochrane Database Syst Rev 2010;CD008475.
91. Selak V, Farquhar C, Prentice A, et al. Danazol for pelvic pain associated with endometriosis. Cochrane Database Syst Rev. 2007;CD000068.
92. Johnson NP, Hummelshoj L; Consensus on current management of endometriosis. World Endometriosis Society Montpellier Consortium. Hum Reprod. 2013;28:1552–68.

Chapter 10
Progestogens and Breast Cancer

Eitan Pe'er

1 Introduction

Progesterone plays a crucial role in the development of the breast alveolar tissue, and ductal branching.

When pregnancy takes place, the placenta produces progesterone in increasing amounts to support the pregnancy and prepare the breast for lactation. As a woman ages, the amount of progesterone produced by the corpus luteum slowly decreases. The clinical results are shortening of the luteal phase, with concomitant shortening of the menstrual cycle and increasing incidence of premenstrual syndrome (PMS) [1].

Progesterone and progestogens are have many uses in gynecology:- including treatment of menstrual disorders, combined oral contraception, and progesterone only contraception, hormone releasing IUD's (LNG-IUS) PMS, prevention of premature uterine contractions, hormonal support of implantation after assisted reproduction, prevention of miscarrage etc [2].

The use of progestogens in menopausal hormone therapy (MHT) gained popularity after Ziel and Finkle's [3] report showing that treatment with estrogen alone causes endometrial hyperplasia and cancer. However, these effects were completely counteracted by the addition of progestone to the estrogen replacement therapy. As hormone replacement therapy (HRT gained popularity, many synthetic forms of progestogen were introduced and are commonly used nowadays). These synthetic progestogens differ from one another in their structure, receptor affinity, metabolism and biological effects as described in other chapters.

E. Pe'er, M.D. (✉)
Gyneco-Endocrinology Unit, Rambam Health Campus,
Technion School of Medicine, Haifa, Israel
e-mail: eitanpeer@gmail.com

H.J.A. Carp (ed.), *Progestogens in Obstetrics and Gynecology*,
DOI 10.1007/978-3-319-14385-9_10

The Women's Health Initiative (WHI) study examined post menopausal hormone replacement therapy with conjugated equine estrogens (CEE) and medroxyprogesterone acetate (MPA). The study was stopped prematurely because of an increased risk of breast cancer (HR-1.26). However in women treated with CEE alone where MPA was not added due to previous hysterectomy there was no increase in the breast cancer risk compared to placebo (HR-0.80) [4]. The lower incidence of breast cancer was also seen in the WHI extended follow-up up study of 11.8 years [5]. Since the publication of the WHI trial there has been additional epidemiological data, all confirming that the addition of progestins to estrogen increases the risk of breast cancer when compared to estrogen treatment alone. Jick et al. [6] reported a case control study of 33,000 women. The reported odds ratio was not significant for developing invasive breast cancer in women using estrogen alone (OR 0.96 95 % CI 0.88–1.06) versus a significantly increased odds ratio of 1.44 (95 % CI 1.31–1.58); for those women using estrogen and progestogen. A meta-analysis [7] of four randomized trials published up to 2006, found a non significant RR of 0.79 (95 % CI = 0.61–1.02) for invasive breast cancer in estrogen only users and a statistically significant increased RR of 1.24 (95 % CI = 1.03–1.50) with estrogen and progestogen. Epidemiological studies have reported a non significant RR of 1.18 (95 % CI = 1.01–1.38) in estrogen only users, and a significantly increased RR of 1.70 (95 % CI = 1.36–2.17) with the use of estrogen plus progestin [7]. In view of the accumulated data cited above, some of the original authors of the WHI report have reiterated their original conclusions, stating that the use of "estrogen-alone treatment" reduces the breast cancer risk and does not substantially interfere with breast cancer detection by mammography [8].

As the molecular understanding of hormonal actions has expanded, with increased clarification of the different modes of action with specific co-activators and co-regulators when attached to their specific receptors, it became clear that not all the progestogens are the same. The French E3N Study comprised of 80,377 postmenopausal women followed up for more than 8 years, using different regimens of hormone therapy. The use of transdermal estrogen with a dehydroprogesterone combination was not associated with an increased risk of breast cancer. All other forms of estrogen progestogen combinations showed an increased breast cancer risk [9].

From a practical and clinical point of view, each progestogen has its own unique biological profile which may, or may not, be shared with other progestogens. The different actions of progestogens are fully described in Chap. 2 on the Pharmacology of the progestogens.

The effect of the normal menstrual cycle on the breast is well known. At the mid-luteal phase, corresponding to the peak level of luteal progesterone, mammary epithelial DNA synthesis and mitosis is at its highest. Hence, progesterone can be clearly seen to have a proliferative effect on the breast epithelium.

During pregnancy, placentally derived high levels of progesterone induce further lobular-alveolar development of the breast, in preparation for lactation. However, if high doses of progesterone are administered before breast surgery, there is less mitotic activity than after treatment with estrogen and progesterone.

Therefore, estrogen and progesterone seem to act synergistically in the breast (and not antagonistically as in the endometrium). It has been shown that a single pulse of progesterone has growth stimulatory effect on cultured breast cells, but subsequent effects of further pulses of progesterone are growth inhibitors. There is an arrest of further cell cycles in the late G1stage. Several factors, such as cyclins D1 D3, decrease, or become inactivated by hypophosphorylation and inhibitory factors (such as kinase inhibitors p21, p27) which are induced by progestins. Hence, there is failure to induce further mitosis by frequent progesterone pulses [10].

2 Progesterone Receptors

The progesterone receptor is a single gene expressed in two isoforms: PR-A (94 kDa) and PR-B (116 kDa). There are homodimers (A-A), (B-B) and heterodimers (A-B) occurring naturally, in different proportions in different target tissues in the female body. They exhibit distinct transcriptional regulatory functions targeting various subsets of genes [11]. The inactivated receptor is activated by hormone (ligand) binding. This hormone-receptor complex translocates to the nucleus where it binds to specific DNA sequences in the promoter regions of target genes to activate gene expression. Alternatively, the expression of specific target genes can be repressed through interaction with various transcriptional factors such as nuclear factor kappa β (NFkβ). Consequently, the clinical significance of activation of the receptor is critically dependent on the transcriptional co-activators and repressors [12]. PR-A is mainly located in the nucleus. This receptor is required for uterine development. PR-B, which is essential for breast development, continuously shuttles between nuclear and cytoplasmatic compartments [13, 14]. Both receptors are co-expressed in the same tissues usually in equal ratios. In humans, normal mammary gland function may rely upon the balanced expression of the two PR isoforms. In breast cancer cells, however this ratio is often altered. An imbalance between the two isoforms appears to be linked to different cancerous phenotypes in the breast [15]. PR-A receptors are more stable and less active. PR-B receptors undergo extensive cross-talk with mitogenic protein kinases and are therefore heavily phosphorylated (more often via action of growth factors) [16]. The PRs can be phosphorylated (and therefore modified) after ligand treatment in response to local growth factors. Mitogenic protein kinase activity is very high in a cancerous environment. They have been shown to persistently phosphorylate, and thus modify PRs action, even in the absence of a ligand. This causes an inappropriate activation of PRs affecting PR modifying binding partners. Therefore PR action differs in normal breast tissue compared to neoplastic breast tissue. Additionally, the presence or absence of estrogen has a significant modifying effect on the PR. These effects are also organ specific, proliferating in the breast inhibitory in the endometrium. In summary, the PR has complex and versatile actions which are:—(1) different in normal and neoplastic tissue. (2) Tissue specific. (3) have different biological clinical actions regarding PR-A and PR-B. (4) May be ER dependent or ER independent. (5) Can act independently without binding ligand (progesterone). (6) different and specific for each progestogen. (7) different if given un a continuous or cyclic fashion. The above effects clearly indicate the varied effects that progestogens have on breast cancer cells biology, but the mechanisms are still not fully elucidated [17].

3 Breast Cancer: The Progesterone Effect

Breast cancer is the most common tumor in women. The life time risk of developing breast cancer is 12 %. Most of the neoplasms present after the menopause and the incidence rises sharply with age. Approximately 5–8 % of tumors arise due to a hereditary predisposition e:g, BRCA-1 mutation. Most breast cancers are initially hormone dependent with estrogen playing a crucial role in development and progression. Breast cancer may develop very slowly having a mean doubling time of 90 days. After a period that may last several years, many breast cancers become hormone independent [18]. Hormone independence may be due a mutation in the estrogen receptor.

Breast cancers are subdivided into two major subtypes:- 1.) Luminal, estrogen receptor positive (ER+), progesterone receptor positive (PR+), cytokeratin 18 positive (CK18+). These tumors make up approximately 80 % of all breast cancers. 2.) Basal ER-, PR-, CK5+. These tumors have a much worse clinical prognosis [19, 20]. In the human breast, normal stem cells are found. They can be defined by expression of epithelial specific antigen (ESA) and α6-integrin. Some believe that ER+ /PR+ luminal cancer cells arise from distinct ER+/PR+ stem cells [21]. However, in recent years it was discovered that in solid ER+/PR+ experimental tumors, CK5+ cells persist that are ER-/PR- stem cells that are actually up regulated by progestins [22]. These colonies, though small in size (up to 100 or less cells) expand and can differentiate into the more common ER+/PR+/CK5-,tumors. Approximately, 2 % of cells in these receptor positive breast cancers retain the ER-/PR-/CK5- stem like signature. [23]. Treatment of ER-/PR-/CK5+ colonies with progesterone, but not with estrogen, leads to an increase in the ER-/PR-/CK5+ stem cell subpopulation from 1 to 2 % to over 20 %. MPA (medroxy-progesterone acetate) has been found to have a more significant effect than other progestins. This increase is independent of estrogen. Hence, it seems that tumor cells can regress from a differentiated state to a more stem-like state, in response to progestins. [24]. Reactivation of the CK5+ cells (with strong CK18 expression, proving it to be of luminal origin) by 24 h of MPA exposure is usually more pronounced in the younger recently formed tumors.

It can be thus concluded that progestins, acting via their progesterone receptor affect the more abundant ER+/PR+/CK5 differentiated cells to reactivate ER-/PR-/CK5+ stem cells. It seems that progestogens have the ability to restore cancer stem cell like properties to some ER+/PR+ breast cancer cells. Only after this effect takes place, can estrogen resume growth and proliferation of the stem cells to expand ER+/PR+/CK5- breast tumors.

4 Local Production and Action of Progestins

The pharmacological properties of progestogens vary according to the parent molecule from which they are derived (testosterone, progesterone, aldosterone etc.) This, together with their ability to activate or deactivate related glucocorticoid receptors, mineralocorticoid receptors and androgen receptors is the reason for the different biological activities specific to each progestogen [25]. The enzyme, 5α-reductase, converts progesterone to 5α-pregnanes (e.g. 5α-dihydroprogesterone= 5α-P). 5α-reductase is highly expressed in cancerous cells but not in normal breast cells. 5α-P which is mainly produced by the cancerous cells stimulates cell proliferation and metastasis via activation of the MAP-kinase pathway [26]. In contrast, in normal breast cells, 5α-reductase activity is low. However, other enzymes, namely 3α-hydroxysteroid oxidoreductase and 20α-hydroxysteroid oxidoreductase metabolize progesterone to 3α-dihydroprogesterone (3α-HP) and 20α-dihydroprogesterone (20α-HP). These metabolites exert an anti-mitogenic effect and increase apoptosis and cell adhesion [27]. In tumor tissue, the concentration of mitogenic 5α-pregnane was 14 times that of the anti mitogenic 3α-HP. In contrast, in the adjacent normal breast cells, the concentration of antimitogenic 3α-HP was more than three times that of the mitogenic 5α-pregnane [26]. In view of the fact that 5α-P stimulates cell proliferation and metastasis, while 3α-HP and 20α-HP exert an antimitogenic effect, increasing apoptosis and cell adhesion, treatment with progesterone may stimulate cell growth only in the presence of malignant breast tissue [28].

Both ER/PR positive and ER/PR negative breast tumors are able to convert progesterone to 3α-HP. Specific high affinity membrane receptors exists for both 5α-P and 3α-HP. These receptors are completely distinct not only from each other but also from known ER, PR, AR and corticosteroid receptors [26]. Levels of 5α-P receptors are up-regulated by 5α-P and estradiol, and down-regulated by 3α-HP in both ER/PR positive and negative tumor cells [28]. 5α-P and 3α-HP have opposing effects on initiation and growth of ER/PR negative human breast tumors. These metabolites, which are independly produced by breast cancer cells, underscore the importance of the microenvironment in regulating expression of receptors, adhesion molecules, growth promoters and inhibitors within the breast cancer cells. 3α-HP maintains normal breast tissue. It inhibits proliferation, tumor initiation and tumor growth. 5α-P has exactly the opposite effects. It induces proliferation, tumor growth and detachment and reduces apoptosis.

5 Progesterone and Migration of Breast Cancer Cells

Death from breast cancer is usually due to invasion and metastasis to distant target organs (e.g. brain, bone lungs). Progesterone and progestins stimulate breast cancer migration and invasion [29]. The migration and invasion of cancer cells involve complex mechanisms. The first step is reorganization of the cells actin/myosin cytoskeleton. This reorganization enables the cells to develop membrane protrusion (called filopodia and lamellipodia). These protrusions are involved in the cell's ability to detach itself from the main tumor body and, to enter lymphatic vessels and migrate. There are several 17β-estradiol up-regulating families of actin-binding protein which leads to this phenomena [30]. Studies comparing the specific effect of different synthetic progestins have demonstrated their individual and different effect on PR+ breast cancer cell migration and invasion in vitro. This activation is mediated in synergy with the activation of the actin-binding protein moesin to increase the formation of membranous structures. These specialized structures interact with the extra cellular matrix of nearby cells, allowing locomotion of the breast cells. [31]. It has been suggested that an in vitro "invasion index" for the effect of specific progestins on breast cancer cells lines incubated in vitro, with the addition of each progestin, with or without estrogen in the media can indicate the different and variable effects of progestins [32].

6 Protease Activated Receptors and Progesterone

Focal Adhesion Kinase (FAK) is a protein that belongs to the family of cytoplasmic tyrosine kinases. When activated, FAK associates with FA (focal adhesion) proteins allowing autophosphorylation at the docking site of the cells. Mature FA complexes, together with phosphorylated FAK recruit other proteins to allow cell detachment [33]. The protease-activated receptors (PARs) are a family of four vascular receptors that respond to local changes in the proteolytic environment. These receptors are activated by thrombin. PARs are important in tissue repair and response to injuries [34]. In tumor cells, PAR2 is activated by factor VIIa-tissue factor (TF), regulating proangiogenic growth factor expression. TF is strongly induced by progesterone in breast cancer cell lines [35].

The ability of progestins to regulate PAR expression appears to be cell specific (endometrium, cervix, vascular vessels, breast etc.). This action is also progestin's specific; levonorgestrel down regulates PAR1 in the endometrium, MPA up regulates PAR1expresion in vascular endothelial cells etc. [33]. Over expression of PAR1 is a feature of many metastatic cancers. These cancers are more invasive *in vitro* [34]. It has been demonstrated that progesterone regulates PAR1 at the mRNA and protein levels [35]. This regulation is dependent on the presence of PR. Advanced breast cancer is associated with a hypercoagulable state. Progesterone has been shown to enhance the coagulation cascade proteins TF and PAR1 promoting cancer cells angiogenesis, coagulation, migration and invasion. This may be another mechanism by which progesterone may contribute to the increase in breast cancer incidence in women using continuous combined HRT.

7 RANK/RANKL and Progesterone

RANKL—receptor activator of the NF-kβ ligand is highly expressed in human breast cancer cells. RANK and its ligand RANKL are expressed in preinvasive mammary intraepithelial neoplasia and invasive carcinoma of the human breast. The dividing epithelial cells do not contain ER or PR, but are controlled by adjacent resting cells containing sex steroid receptors which secrete various growth factors, of which RANKL emerges as a key paracrine mediator of the progesterone mitogenic signal. It has been shown that progesterone-knock-out (PRKO) mice mammary epithelial cells, when mixed with wild-type (WT) cells, contribute to alveolar and ductal side branching in the pregnant state. This suggests a paracrine factor transmitted from the wild-type breast cells and received by the knock-out cells causing them to proliferate. This paracrine factor was found to be RANKL [36].

Mammary RANKL is induced by exogenous progesterone. In the proliferative phase of pregnancy (in which progesterone is at its peak action driving mammary epithelium expansion and morphogenesis), RANKL is markedly expressed. RANKL expression is confined to ER+/PR+ transmitters' cells [37]. Consequently RANKL may act as the direct link between breast cells via progesterone. Finally, evidence that RANKL has a mediator role in mammary progesterone signaling, came from a study where PRKO mammary epithelial cells were transplanted into the mammary fat pad of WT mice. RANKL triggered mammary side-branching and alveolar budding in the PRKO transplant within the pregnant WT host [38]. It has been demonstrated that the RANKL transduction axis is actually essential for progesterone promotion of mammary tumorigenesis [39, 40]. In these studies MPA (medroxyprogesterone acetate), the progestin used in the WHI study, significantly increased RANKL expression in the ER+/PR+ cell population, in mammary normal epithelium as well as in premalignant and malignant cells. Therefore it is evident that progesterone (as well as MPA) relies on RANKL as a paracrine mediator for its proliferative effects. RANKL also enables the mammary epithelium to evade premature apoptosis.

8 Progesterone and E-Cadherin

E-Cadherin is an epithelial adhesion protein, which is an important, if not the major, component of the tight junctions between mammary epithelial cells. It has been shown that decrease, or loss, of the E-Cadherin protein is associated with tumor cell metastasis and invasiveness, and a poorer prognosis in breast cancer patients [41]. E-cadherin protein is highly expressed in normal epithelial cells adjacent to the breast tumors. In an experimental rat model, treatment with E + the R5020 (promegestrone) decreased the levels of E-cadherin precursor and mature E-cadherin protein [42]. This effect was abolished by the progesterone antagonist mifepristone (RU486), implying that the effect is due to the progestin component. In this rat model E+P treatment, as compared with E alone resulted in invasive mammary cancers accompanied by decreased E-cadherin levels and expansion of cells with a basal/myoepithelial phenotype. Similar findings have been observed in invasive primary human breast cancers compared to matched carcinoma in situ. While estrogen alone is sufficient to induce luminal noninvasive tumors, progesterone is required for the expansion of basal-myoepithelial tumor cells that frequently express progesterone receptor B. Progesterone promotes expansion of the more invasive basal/myoepithelial cells via direct activation of progesterone receptor B. It is important to note that only progesterone receptor B mediates the effect of progestins on E-cadherin. The ratio of progesterone receptors A/B is almost equal in normal breast tissue. But this ratio is completely altered, in favor of more B receptor in cancerous breast cells.

Loss of E-cadherin is a hallmark of the invasive behavior of luminal breast cancer cell lines. In vitro studies have found that treatment with E+P leads to complete loss of E-cadherin in 77 % of the tumors in the rat model. The data now indicate that Progesterone acting through progesterone receptor B regulates E-cadherin protein expression. Progestins, therefore specifically contributes to the aggressiveness and invasion of breast cancer cells.

9 Conclusion

Clinical studies, such as the WHI, Million women study and others have accelerated the research to find "evidence" to prove, or disprove, the clinical findings. It soon became apparent that progesterone, the *"insignificant hormone"* partner of estrogen in hormone replacement therapy plays an important and complex role in mammary carcinogenesis. Progesterone works synergistically with estrogen on the breast, inducing alveolar proliferation. Progesterone has always been considered as an antiproliferative hormone due to its action on the endometrium, opposing the proliferative effect of estrogen. Hence progestogens were introduced into hormone replacement regimens: to counteract estrogen induced endometrial proliferation. Many laboratories throughout the world are trying to elucidate progesterone dependent mechanisms, in order to clear controversial issues reported in clinical studies. However, thus far it is clear that there is much more to progestogens and their widespread actions on genes, receptors and tissues causing, promoting and inducing breast cancer. In view of the new information gathered so far, new well designed prospective clinical studies are now needed.

References

1. Brown JB. Types of ovarian activity in women and their significance. Hum Reprod Update. 2011;17(2):141–58.
2. Ringa V, Fritel X, et al. Association between premenstrual progesterone and postmenopausal hormone therapy in the GAZEL cohort. Maturitas. 2007;58:216–25.
3. Ziel HK, Finkle WD. Increased risk of endometrial carcinoma among users of conjugated estrogens. N Engl J Med. 1975;293(23):1167–70.
4. Chelbowski RT, Hendrix SL, et al. Influence of estrogen plus progestin on breast cancer and mammography in healthy postmenopausal women;- The women health initiative randomized trial. JAMA. 2003;289:3242–53.
5. Anderson GL, Chlebowski RT, et al. Conjugated equine oestrogen and breast cancer incidence and mortality in postmenopausal women with hysterectomy; extended follow up of the women's health initiative randomized placebo-controlled trial. Lancet Oncol. 2012;13:476–86.
6. Jick SS, Hagberg KW, et al. Postmenopausal estrogen-containing hormone therapy and the risk of breast cancer. Obstet Gynecol. 2009;113:74–80.
7. Collins JA, Blake JM. Breast cancer risk with postmenopausal hormone treatment. Reprod Update. 2006;121:331.
8. Chlebowski RT, Anderson GL. Changing concepts: menopausal hormone therapy and breast cancer. J Natl Cancer Inst. 2012;1041:517–27.
9. Fournier A, Berrino F, et al. Breast cancer risk in relation to different types of hormone replacement therapy in the E3N-EPIC cohort. Int J Cancer. 2005;114:448–54.
10. Gaham JD, Yager M, et al. Altered progesterone receptor isoforms expression remodels progestins responsiveness of breast cancer cells. Mol Endocrinol. 2005;19:2713–35.
11. Khan JA, Bellance C, et al. Differential regulation of breast cancer-associated genes by progesterone receptors isoforms PRA and PRB in a new bi-inducible breast cancer cell line. PLoS One. 2012;7:45993.
12. McKenna NJ, Lanz RB, et al. Nuclear receptor coregulators: cellular and molecular biology. Endocr Rev. 1999;20:321–44.
13. Guiochon-Mantel A, Delabre K, et al. Nuclear localization signals also mediate the outward movement of proteins from the nucleus. Proc Natl Acad Sci U S A. 1994;91:7179–83.
14. Boonyaratanakornkit V, Bi Y, et al. The role and mechanism of progesterone receptors activation of extra-nuclear signaling pathways in regulating gene transcription and cell cycle progression. Steroids. 2008;73:922–8.
15. Mote PA, Bartow S, et al. Loss of co-ordinate expression of progesterone receptors A and B is an early event in breast carcinogenesis. Breast Cancer Res Treat. 2002;72:163–72.
16. Clemm DI, Sherma L, et al. Differential hormone-dependent phosphorylation of progesterone receptor A and B forms revealed by phosphoserine site-specific monoclonal antibody. Mol Endocrinol. 2000;14:52–65.
17. Hagan CR, Lange CA. Molecular determinants of context-dependent progesterone receptor action in breast cancer-minireview. BMC Med. 2014;12:32.
18. Pasqualini JR. Breast cancer and steroid metabolizing enzymes: the role of progesterone. Maturitas. 2009;655:s17–21.
19. Perou CM, Sorlie T, et al. Molecular portraits of human breast tumors. Nature. 2000;406:747–52.
20. Al-Hajj M, Wicha MS, et al. Prospective identification of tumorigenic breast cancer cells. Proc Natl Acad Sci U S A. 2003;100:3983–8.
21. Dontu G, El-Ashry D, et al. Breast cancer/progenitor cells and estrogen receptor. Trends Endocrinol Metab. 2004;15:193–7.
22. Clarke RB. Ovarian steroids and the human breast, regulation of stem cells and cell proliferation. Maturitas. 2006;54:327–34.

23. Horwitz KB, Dye WW, et al. Rare steroid receptor-negative basal-like tumorigenic cells in luminal subtype human breast cancer xenografts. Proc Natl Acad Sci USA. 2008;105:5774–9.
24. Horwitz HB, Sartorius CA. Progestins in hormone replacement therapies. Reactive cancer stem cells in women with pre-existing breast cancer-A hypothesis. J Clin Endocr Metab. 2008;93(9):3295–8.
25. Stanczyk FZ, Hapgood JP, et al. Progestogens used in postmenopausal hormone therapy. Differences in their pharmacological properties, intracellular actions and clinical effect. Endocr Rev. 2013;34(2):171–208.
26. Kuhl H, Schneider HPG. Progesterone- a promoter or inhibitor of breast cancer. Climacteric. 2013;16 suppl 1:54–68.
27. Wiebe JP, Beausoleil M, et al. Opposing actions of progesterone metabolites 5α-dihydroprogesterone (5αP) and 3α- Bcl-2, Bax and p21 in human breast cell lines. J Steroid Biochem Mol Biol. 2010;118:125–32.
28. Wiebe JP. Progesterone metabolites in breast cancer. Endocr Relat Canc. 2006;13:717–38.
29. Kato S, Pinto M, et al. Progesterone increases tissue factor gene expression, procoagulant activity, and invasion in the breast cancer cell line ZR-75-1. J Clin Endocr Metab. 2005;90:1181–8.
30. Polland TD, Borisy GG. Cellular motility driven by assembly and disassembly of filaments. Cell. 2003;112(4):453–60.
31. Fu XD, Giretti MS, et al. Comparative actions of progesterone, medroxyprogesterone acetate, drospirenone and nestorone on breast cell migration and invasion. BMC Cancer. 2008;8:166. 1–14.
32. Critchley DR. Focal adhesions- the cytoskeletal connection. Curr Opin Cell Biol. 2000;12:133–9.
33. Ossovskaya VS, Bunnett NW. Protease-activated receptors; contribution to physiology and disease. Physiol Rev. 2004;84:579–621.
34. Hague S, Oehler MK, et al. Protease activated receptor-1 is down regulated by levonorgestrel in endometrial stromal cells. Angiogenesis. 2002;5:93–8.
35. Even-Ram S, Uziely B, et al. Thrombin receptor overexpression in malignant and physiological invasion processes. Nat Med. 1998;4:909–14.
36. Diaz J, Aranda E, et al. Progesterone promotes focal adhesion formation and migration in breast cancer cells through induction of protease-activated receptor -1. J Endocrin. 2012;214:165–75.
37. Fernandez-Valdivia R, Lydon JP. From the ranks of mammary progesterone mediators, RANKL takes the spotlight. Mol Cell Endocrinol. 2012;357(1–2):91–100.
38. Fernandez-Valdivia R, Mukherjee A, et al. The RANKL signaling axis is sufficient to elicit ductal side-branching and alveologenesis in the mammary gland of the virgin mouse. Dev Biol. 2009;328:127–39.
39. Beleut M, Rajaram RD, et al. Two distinct mechanisms underlie progesterone-induced proliferation in the mammary gland. Proc Natl Acad Sci U S A. 2010;107:2989–94.
40. Gonzalez-Suarez E, Jacob AP, et al. RANK ligand mediates progestin-induced mammary epithelial proliferation and carcinogenesis. Nature. 2010;468:103–7.
41. Schramek D, Leibbrandt A, et al. Osteoclast differentiation factor RANKL controls development of progestin-driven mammary cancer. Nature. 2010;468:98–102.
42. Siitonen SM, Kononen JT, et al. Reduced E-cadherin expression is associated with invasiveness and unfavorable prognosis in breast cancer. Am J Clin Pathol. 1996;105:394–402.

Chapter 11
Progestogens in Endometrial Cancer

Ido Laskov and Walter H. Gotlieb

1 Introduction

Endometrial cancer is the most common gynecologic malignancy in affluent countries [1]. Approximately 15 % of women will be diagnosed with endometrial cancer before menopause, and 4 % will develop the disease before the age of 40 years [2, 3]. The current therapeutic approach for early-stage endometrial cancer includes total abdominal hysterectomy with bilateral salpingo-oophorectomy, and lymphadenectomy (pelvic and aortic), depending on preoperative or intraoperative pathologic risk profiles. Women with grade 1 endometrial cancer without myometrial invasion (consistent with FIGO 1988 stage Ia) treated by conventional surgery have a disease-specific survival of 99.2 % after 5 years, and 98 % after 10 years [4]. Many endometrial cancer patients carry an increased burden of medical co-morbidities, such as obesity, diabetes mellitus and hypertension, conditions that have been associated with a higher risk of surgical adverse events [5, 6]. Some patients carry severe medical co-morbidities that may preclude them having surgery [7, 8]. This high risk group of patients with endometrial cancer may sometimes have received progestin treatment as an alternative to surgery, or as treatment until they are deemed as fit for surgery.

I. Laskov, M.D.
Division of Gynecologic Oncology, Jewish General Hospital & Lady Davis Research Institute, McGill University, Montreal, QC, Canada

W.H. Gotlieb, M.D., Ph.D. (✉)
Division of Gynecologic Oncology, Jewish General Hospital, McGill University, 3755 Cote Ste Catherine Road, Room E716, Montreal, QC, Canada H3T1E2
e-mail: walter.gotlieb@mcgill.ca

H.J.A. Carp (ed.), *Progestogens in Obstetrics and Gynecology*,
DOI 10.1007/978-3-319-14385-9_11

A second group of patients that might benefit from progestin treatment are young women diagnosed with endometrial cancer who still wish to have children. The excellent cure rates that are attained for well differentiated endometrial cancer have allowed us to shift the focus, from the high survival towards quality of life issues after treatment. In particular, there has been increased attention focused on fertility preservation, as approximately one of ten patients with endometrial cancer develop the disease during reproductive age. Considering that the average age at first birth has steadily increased in developed nations. First births are now 1 in 12 births in women above the age of 35 (compared to only 1 in 100 in 1970) [9]. Hence, it is not surprising that many of these younger women with endometrial cancer desire fertility preserving options. The decision to proceed with conservative management is complex. Many issues related to fertility-preservation for endometrial cancer remain uncertain, and warrant further consideration. The optimal work-up to evaluate the extent of disease in young patients with endometrial cancer who desire to maintain their uterus remains unclear, as is the lack of uniformity in the medical management and surveillance.

In this chapter we will discuss the role of progestin hormonal therapy in the conservative management of endometrial cancer, including the potential risks associated with medical management compared to surgical care, the appropriate candidate selection and work up, the expected outcomes, the variety of progestational agents that have been used, and the recommended follow-up (Fig 11.1). We will also describe the role of progestins in advanced endometrial cancer and the foreseeable future of progesterone treatments in endometrial cancer.

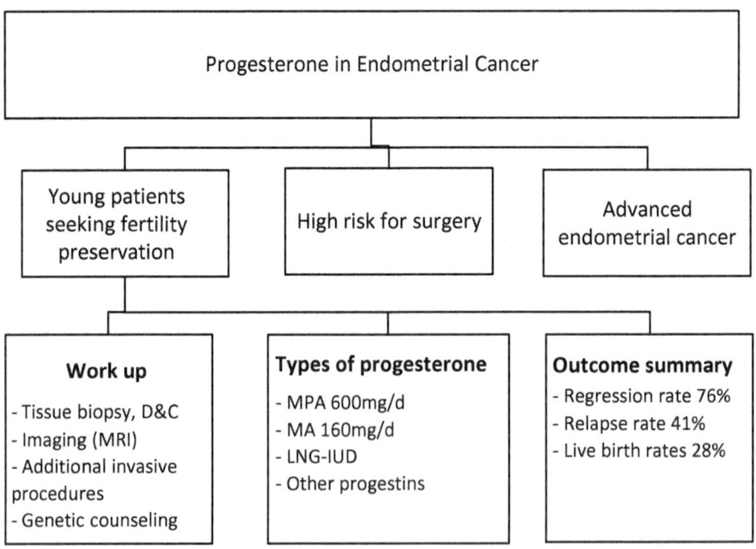

Fig. 11.1 Progesterone in endometrial cancer, chapter summary. *MPA* medroxyprogesterone acetate, *MA* megesterol acetate, *LNG-IUD* levonorgestrel releasing intrauterine device

2 Progesterone for Fertility Preservation

2.1 *Are There Any Risks?*

The development of endometrial cancer in young women usually results from a hyperestrogenic state that leads to endometrial hyperplasia. A tissue biopsy of atypical endometrial hyperplasia has been associated with a 29 % risk of progression to endometrial cancer [10], and endometrial cancers have been found in up to 43 % of patients with a preoperative diagnosis of atypical endometrial hyperplasia [11]. This high association warrants consideration in management decisions. According to a review of over 2,000 women aged 40 years or younger collected from the National Cancer Institute database, the majority of patients (75 %) had disease confined to the uterus, but approximately 17 % had stage III or IV disease [10]. These younger patients are also at increased risk of other pathological gynecologic conditions, including ovarian tumors. In a review of young women with endometrial cancer by Walsh et al. [11], 26 of 102 women (25 %) were found to have coexisting epithelial ovarian tumors (23 synchronous primaries and 3 metastases). Therefore, any decision to deviate from the standard approach of hysterectomy with oophorectomy and staging should take account of the risk of an undetected, and therefore subsequently untreated, synchronous or metastatic cancer. These studies confirm the need for thorough examination and careful patient selection, while highlighting the risks inherent in conservative management of an unstaged cancer.

Based on these data, the patient's outcome may be adversely affected when choosing to pursue fertility preservation. In the absence of randomized trials, the largest study to evaluate this matter is a retrospective study by Koskas et al. [12] who examined 489 patients aged 40 or younger with grade 1 endometrial adenocarcinoma. The patients were divided into groups who underwent uterine preservation, ovarian preservation, or hysterectomy with oophorectomy. Ovarian and uterine preservation had no effect on either cancer-specific or overall survival. The limitations of Koskas et al's. [12] study include the absence of information on which agents and treatment protocols were used and how they found no evidence for the 17–25 % of young patients with concomitant/metastatic adnexal carcinomas published in other reports [12, 13].

2.2 *Workup Prior to Treatment*

The optimal work-up to evaluate the extent of disease in young patients with endometrial cancer who desire to maintain their uterus has not been established. Every effort should be taken to ensure that the cancer is confined to the endometrium and low grade, and therefore likely to respond to hormonal therapy without compromising curability (Table 11.1). As a rule, the pretreatment evaluation should consist of a full workup for any signs or symptoms suspicious for advanced/metastatic disease (Tables 11.2 and 11.3).

Table 11.1 Suggested criteria for progesterone treatment [39]

I	Absence of frank myometrial invasion
II	Well-differentiated (G1) endometrioid adenocarcinoma
III	No contraindications for progesterone therapy
IV	Potential for fertility
V	Informed consent on the indications and limitations of progesterone therapy

Table 11.2 Suggested procedures for the assessment of a patient with endometrial cancer seeking fertility sparing treatments

Procedure	Purpose
Complete history and physical exam	Look for signs or symptoms suspicious for advanced/ metastatic disease
D&C	– Tumor grading – Possible therapeutic effect
MRI	Assess myometrial invasion and loco-regional disease spread
Diagnostic laparoscopy	Partial surgical staging
Sentinel lymph node biopsy	Value to be determined
Genetic counseling	Risk assessment for patient and family

Table 11.3 Fertility sparing options; advantages and disadvantages

Drug	Dose	Advantages	Disadvantages
Medroxyprogesterone acetate	400–600 mg/day for at least 3 months	Well studied	Known side effects
Megesterol acetate	160–320 mg/day for at least 3 months	Well studied	– Known side effects – Might have higher recurrence rate compared to MPA
Progesterone - intrauterine device	20–65 mcg/day	Low systemic toxicity	– Limited data – Intra-uterine placement required
Natural progesterone	200 mg/days, days 14–25	–	Limited data
Hydroxyprogesterone	500 mg/days	–	Limited data
Norethisterone	5 mg/days	–	Limited data
Progestogens at various doses	–	–	Limited data

2.2.1 Tissue Biopsy

Prior to initiating conservative management, dilatation and curettage (D&C) is recommended because it better defines the grade of the tumor compared to office endometrial biopsy [13]. Additionally, there might be value in the removal of most of the endometrial cancer cells by the D&C before starting hormonal treatment [14].

2.2.2 Imaging

Attempts should be made to rule out myometrial invasion, adnexal involvement and lymph node metastases, which are regarded as contra-indications for conservative management. MRI has proven to be superior to transvaginal ultrasound or CT for determining myometrial invasion [15]. Pooling of 11 studies, comparing T2-weighted imaging and contrast-enhanced magnetic resonance imaging, revealed similar positive predictive values for myometrial invasion of 0.65 and negative predictive values of 0.85 [16]. MRI is used to assess loco-regional disease spread [17], and Sironi et al. [15] reported a sensitivity and specificity of 74 % for MR assessment of superficial myometrial invasion, although the importance of superficial myometrial involvement on response to progestins is not clear.

2.2.3 Additional Invasive Procedures

There is an increased risk of concomitant adnexal involvement in premenopausal patients with endometrial cancer, reaching up to 25 % in the series from Cedars Sinai [11]. Consequently, some physicians perform a diagnostic laparoscopy at the time of D&C [18]. With the evolving data on sentinel lymph node biopsy for endometrial cancer, lymph node biopsy could be considered in selected cases [19].

2.2.4 Genetic Counseling

Women diagnosed with endometrial cancer at a young age are at increased risk for mismatch repair gene mutations associated with Lynch syndrome [20]. Hence, these women should also be referred for genetic counseling [21], as counseling might reveal important implications concerning the risk of adnexal pathology and colon cancer necessitating screening in these young patients and their families.

2.3 Prognostic Factors

Although the majority of carefully selected patients will respond to progestin therapy, there is at present no way to accurately predict who will respond.

Data remain scarce on clinical or pathologic predictors of response to progestin treatment in premenopausal women with complex and Grade 1 endometrial adenocarcinoma. Park et al. analyzed 148 patients (age $\leq$40 years) with stage IA, grade 1, endometrioid adenocarcinoma of the uterus who underwent fertility-sparing management using daily oral medroxyprogesterone acetate or megestrol acetate [24]. One hundred and fifteen (77.7 %) showed complete response to progestin treatment, and 35 (30.4 %) experienced recurrence after a median follow-up period of 66 months. A body mass index (BMI) $\geq$25 was the only significant factor associated with a failure to achieve cure (odds ratio [OR], 3.00; 95 % CI, 1.35–6.66; P=0.007). A BMI ≥25 was also significantly associated with a higher risk of recurrence (OR, 2.14; 95 % CI, 1.06–4.31; P=0.033). The use of MPA (compared to MA) (OR, 0.44; 95 % CI, 0.22–0.88; P=0.021), continuing maintenance treatment (OR, 0.22; 95 % CI, 0.05–0.94; P=0.042), and a previous pregnancy (OR, 0.25; 95 % CI, 0.11–0.56; P=0.001) were significantly associated with a lower risk of recurrence [22]. Penner et al. [23] looked at the histopathologic features, using a qualitative abnormal endometrial architecture score, comparing pretreatment and follow-up endometrial specimens to identify predictors of resolution [25]. The score is composed of five features: polypoid, cribriform, papillary, budding and back to back endometrial glands. Resolution rates expressed as the Standardized Resolution Ratio (SRR) were highest in individuals with a low pre-treatment score and a BMI <35 (SRR=1.48, p=0.03), lower among subjects with a high pre-treatment score (SRR=0.37, p<0.03), and lowest in subjects whose first follow-up specimen showed persistent complexity, atypia, or carcinoma with adjacent stromal decidualization (SRR=0.24, p=0.002) [23]. The presence of progesterone receptors also predicts response to progestin therapy [24, 25]. In one study the response rate was 8 % (seven of 86 patients) for patients who were progesterone receptor-negative and 37 % (17 of 46) for patients who were progesterone receptor-positive (*P*<.001) [26]. In addition, PTEN and KRAS status in combination with the progesterone receptor expression in the tumor appear promising as biomarkers of response [25]. Further investigations in predictors of response may ultimately lead to personalized treatments for young women with endometrial cancer.

2.4 Types of Progesterone

At present, there is no consensus on the optimal medication, dose, or length of treatment. In a 2004 review, the most commonly used agents were medroxyprogesterone acetate 500–600 mg (MPA; 44 %) and megesterol acetate 160 mg (35 %) for at least 3 months [27]. Both regimens appear to have similar response rates although it has recently been suggested that the recurrence rate is higher after megestrol acetate compared to medroxyprogesterone acetate [22].

Additionally, treatment has been reported with the levonorgestrel intrauterine device (LNG-IUD) (Mirena^{TDM}) that releases 20 mcg of levonorgestrel per day [28], in combination with hysteroscopic resection [29], medroxyprogesterone acetate [30] or GnRH analogues [31]. Other treatments used include intramuscular 17-hydroxyprogesterone, oral contraceptive pills, norethisterone, dihydrogesterone, and natural progesterone either utilized alone or in a combination of progestin agents [7, 32].

The choice of progestin should be based on measurable outcomes, including efficacy, side-effects, and patient tolerability. Orally administered progestins are not without side-effects, including mood alterations, headaches, weight gain, breast pain and/or tenderness, and increased risk of thrombus formation. Thrombosis is a serious adverse reaction to MPA. It is caused by the inhibitory activity of MPA against plasminogen activator [33]. Thrombosis can be fatal especially if leading to cerebral infarction, myocardial infarction, or pulmonary embolism. Clotting factors should be checked monthly, and treatment with MPA should be discontinued on detection of clotting abnormalities. A prospective trial using 600 mg MPA [34] reported that the most common side effects were weight gain and liver dysfunction. There were no cases of thromboembolism. Progesterone therapy is contraindicated in the presence of a history of thromboembolus, breast cancer, or hepatic dysfunction. The progesterone-releasing IUD might be a means of achieving a localized effect within the endometrium while avoiding the adverse systemic toxicity. There is no consensus regarding the optimal progestin duration. Progestin therapy has an impact on the endometrial cells as early as 10 weeks after initiation of treatment, but most physicians suggest the need for a minimum of 3 months of treatment before assessing the response with endometrial hyperplasia and even longer for endometrial cancer [34]. Obese and anovulatory women have been shown to require longer periods of progestin therapy to attain a complete response, and are more prone to relapse [22, 23].

2.5 *Outcome*

Although the first publication describing fertility preserving, conservative treatment with hormones was published in 1961 [35], the number of publications describing the outcome is still limited (Fig. 11.2) and many questions remain. The possibility of publication bias in the studies analyzed should be borne in mind. Studies showing treatment success are more likely to be reported and published than negative trials, leading to overestimating the success rate. In a recent metaanalysis [36], including 34 observational studies, the authors evaluated the regression, relapse, and live birth rates of 408 women diagnosed with early-stage endometrial cancer. The primary studies included the outcome of women with well-differentiated endometrial cancer with 386 women being classified as G1 and 22 women with moderate or poor differentiation (G2 or G3). Half of the studies were prospective cohorts (17 of 34) and only in 6 of the 34 studies, was the follow-up more than 5 years. Overall, resolution occurred in 76 % (301/408) of reported patients (Table 11.4), and 89 (40.6 %) responders relapsed during follow-up. 75 women achieved at least 1 live birth, yielding a live birth rate of 28 %.

3.6 % of patients were diagnosed with ovarian malignancy during follow-up. It is unclear whether these represent concurrent ovarian malignancies or metastatic ovarian involvement from the primary endometrial neoplasm. There were also ten women (1.8 %) diagnosed with stage II disease or greater following treatment failure, and there were two deaths reported (0.5 %). Another recent systemic review by Gunderson et al. [32] reported oncologic and reproductive outcomes with progestin therapy in women with endometrial hyperplasia and grade 1 endometrial cancer. Forty-five studies with 391 study subjects were identified including 280

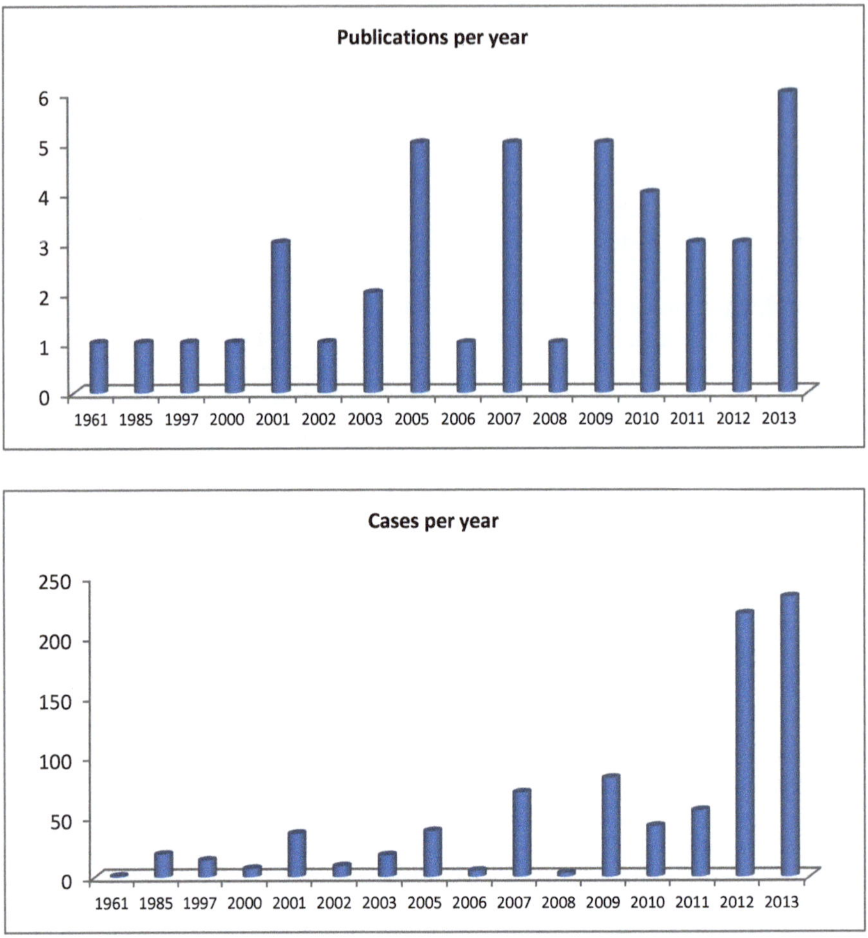

Fig. 11.2 Number of publications and number of reported cases treated conservatively

women that had grade 1 endometrial adenocarcinoma. The median age for the overall cohort was 31.7 years (range 19–80 years). When stratified by disease type, the durable complete response rate was significantly higher in women with complex atypical hyperplasia (65.8 %) compared to those with carcinoma (48.2 %; p = .002). The rate of initial response in women with complex atypical hyperplasia was also significantly higher (85.6 %), than women with carcinoma (74.6 %; p = 0.03). Disease recurrence was more likely to occur in the carcinoma cohort (35.4 %) than the hyperplasia group (23.2 %; p = .03). Further, persistent disease was noted in only 14.4 % of women with complex atypical hyperplasia compared with 25.4 % of those with carcinoma (p = .02). Reproductive outcomes did not differ between the cohorts.

Table 11.4 Overview of studies and outcomes of progestogen treatment, adapted from Gallos et al. [36]

Study (year, reference)	Total of patients	Regressed (%)	Relapsed (%)	Live births (%)
Bokhman (1983, [61])	19	15 (79)	–	–
Randall (1999, [62])	14	10 (71)	1/10 (10)	3/14 (21)
Kim (2000, [63])	7	4 (57)	2/4 (50)	0/7 (0)
Imai (2001, [64])	14	8 (57)	3/8 (38)	2/14 (14)
Kaku (2001, [65])	12	9 (75)	2/9 (22)	1/12 (8)
Duska (2001, [2])	12	9 (75)	–	–
Wang (2002, [66])	9	8 (89)	4/8 (50)	2/9 (22)
Gotlieb (2003, [67])	13	13 (100)	6/13 (46)	3/13 (23)
Jadoul (2003, [68])	5	3 (60)	0/3 (0)	3/5 (60)
Niwa (2005, [69])	12	12 (100)	8/12 (67)	–
Ota (2005, [70])	12	5 (42)	2/5 (40)	2/12 (17)
Yahata (2005, [71])	8	7 (88)	7/7 (100)	2/8 (25)
Yang (2005, [72])	6	4 (67)	2/4 (50)	2/6 (33)
Le Digabel (2006, [73])	5	3 (60)	1/3 (33)	0/5 (0)
Elizur (2007, [74])	8	8 (100)	3/8 (38)	4/8 (50)
Minaguchi (2007, [75])	18	14 (78)	5/14 (36)	1/18 (6)
Ushijima (2007, [34])	22	14 (64)	8/14 (57)	3/22 (14)
Wheeler (2007, [76])	21	7 (33)	1/7 (14)	–
Yamazawa (2007, [77])	9	7 (78)	2/7 (29)	3/9 (33)
Li (2008, [78])	3	3 (100)	0/3 (0)	–
Eftekhar (2009, [79])	21	18 (86)	3/18 (17)	2/21 (10)
Hahn (2009, [80])	35	22 (63)	9/22 (41)	8/35 (23)
Han (2009, [81])	7	7 (100)	0/7 (0)	5/7 (71)
Signorelli (2009, [82])	11	6 (55)	4/6 (67)	4/11 (36)
Yu (2009, [83])	8	6 (75)	1/7 (17)	0/8 (0)
Mao (2010, [84])	6	4 (67)	0/4 (0)	3/6 (50)
Mazzon (2010, [85])	6	6 (100)	0/6 (0)	4/6 (67)
Minig (2010, [31])	14	8 (57)	2/8 (25)	1/14 (7)
Cade (2010, [7])	16	10 (63)	–	–
Laurelli (2011, [29])	14	14 (100)	1/14 (7)	1/14 (7)
Park (2011, [86])	14	13 (93)	3/13 (23)	13/14 (29)
Perri (2011, [87])	27	24 (89)	9/24 (38)	12/27 (44)
Total	408	301 (76 %)	89/267 (40.6 %)	75/325 (28 %)

2.5.1 Repeat Treatment for Recurrence After Complete Response?

Park and colleagues recently published a retrospective multicenter study that shows the safe and effective outcome of re-treating 33 young patients who still wanted to preserve fertility following recurrence after a complete response to progestins [37]. Five of the 33 women failed to respond to a second conservative approach, and another five patients recurred after a second complete response. Three received a third cycle of progestins and two responded again. Five patients delivered six healthy babies following this second conservative approach. The responders were followed for a mean of period of 51 months, no patient died of disease or suffered an adverse outcome.

2.5.2 Outcome for Progestin Releasing Intrauterine Devices

Levonorgestrel releasing intrauterine devices (LNG-IUD) are associated with contraceptive efficacy, powerful reduction of menstrual blood volume through suppression of endometrial growth, and accompanying relief of menstrual pain [38]. It has also been shown that the use of LNG-IUD in combination with hormone replacement therapy during or after menopause can prevent endometrial cancer [39]. The efficacy of the LNG-IUD in suppressing the growth of endometrium has also been shown in patients with "non-atypical" endometrial hyperplasia, with a response seen in 96 % of all cases within 1 year and 92 % within 2 years after insertion [40]. Efficacy in patients with endometrial cancer is presently being investigated. Preliminary data obtained from two separate studies suggests that progestin treatment provided by an IUD in 22 patients with grade 1 Stage I endometrial cancer [29, 41] was followed by a 68 % (15/22) complete response after six months or longer compared to 72 % (73/102) of patients on oral progestin [8]. No relapses or progressions were reported after 6–71 months of follow-up. Fertility outcomes were not reported. In addition, a few more studies, all with a small number of patients [30, 31] suggests that treatment with oral or intrauterine progestin is similarly effective. Large prospective trials for LNG-IUD are presently underway in order to clarify some of the unresolved issues (Table 11.5).

Table 11.5 Progestin and endometrial cancer, ongoing clinical trials and time of expected results

ClinicalTrials.gov Identifier #	Title	Estimated study completion date
NCT00483327	Management of atypical endometrial hyperplasia and endometrial carcinoma using megestrol acetate	October 2013
NCT01594879	Treatment with medroxyprogesterone acetate plus LNG-IUS in young women with early stage endometrial cancer	December 2014
NCT00788671	Levonorgestrel intrauterine device (IUD) to treat complex atypical hyperplasia (CAH) and Grade 1 endometrioid endometrial cancer	November 2015
NCT01943058	Megestrol acetate or levonorgestrel-releasing intrauterine system in treating patients with atypical endometrial hyperplasia or endometrial cancer	October 2016
NCT02035787	Metformin with the levonorgestrel-releasing intrauterine device for the treatment of complex atypical hyperplasia (CAH) and endometrial cancer in non-surgical patients	March 2017
NCT00003179	Surgery plus medroxyprogesterone in preventing endometrial cancer	N/A

2.6 *Follow Up*

In view of the high relapse rates (35–41 % [32, 36]), the frequency of concomitant adnexal malignancy, and the risk of upgrading of the cancer, close follow is essential.

Thinning of the endometrium as seen on transvaginal ultrasound is associated with an increased chance of responding to progestin therapy [34]. However, the predictive value is insufficient to negate endometrial sampling. It is important to note that the diagnostic accuracy of endometrial aspiration biopsy (pipelle) while the LNG-IUD is in place may not be as accurate as dilatation & curettage (D&C) [42].

It is sensible to recommend staging hysterectomy with bilateral salpingo-oophorectomy once the family is complete or if fertility-sparing treatment fails, either due to failure of regression or relapse. When regression occurs, some recommend assisted reproduction to maximize the chances of a live birth and decrease the time to definitive treatment. Additionally, immediate assisted reproduction avoids prolonged unopposed estrogen stimulation, which could cause relapse.

The need for oophorectomy together with hysterectomy remains debateable in view of the risk of concomitant ovarian involvement [11]. One series based on the database from the National Cancer Institute's Surveillance, Epidemiology, and End Results (SEER) suggested that there was no increase in cancer related mortality associated with ovarian preservation in young women with early stage endometrial cancer [43].

3 Progesterone in Advanced Endometrial Cancer

Progesterone has long been used for advanced or recurrent endometrial cancer. Kauppila [44] reviewed 1,068 patients treated with medroxyprogesterone acetate (MPA), megestrol acetate, or hydroxyprogesterone caproate in different trials, and found an overall response rate of 34 %, with a mean duration of response ranging from 16 to 28 months and a mean survival ranging from 18 to 33 months. However subsequent clinical studies, based on more stringent criteria for response assessment, reported lower response rates ranging from 11 to 16 %, without any significant difference according to the type of progestin used [45, 46]. When MPA 200 or 1,000 mg/day was administered to 229 patients with advanced or recurrent endometrial cancer, the low-dose group (200 mg/day group) showed better outcomes (17 % complete remission and 8 % partial remission) than the high-dose group (1,000 mg/day group) (9 % complete remission and 6 % partial remission) [26]. Median progression free survival and overall survival rates were 3.2 months and 11.1 months, respectively, for the low-dose group, and 2.5 and 7 months for the high dose group, with no differences in toxicity noted between the two arms. This GOG trial showed that the response rate was higher in the low-dose group than in the high-dose group with the responses being particularly favorable in G1 and PR-positive cases [26]. The combination of progesterone therapy and chemotherapy has been tested in small

Table 11.6 Hormonal therapy with progestins and tamoxifen in advanced or recurrent endometrial cancer

Study	Hormonal agent	Patients	CR (%)	PR (%)	OR (%)
Thigpen JT [26]	MPA (200 mg/day)	145	17	8	25
	MPA (1,000 mg/day)	154	9	6	15
Thigpen JT [88]	TAM	68	4	6	10
Pandya KJ [49]	MA	20	5	15	20
	TAM/MA	42	2	17	19
Whitney CW [47]	TAM/MPA	58	10	23	33
Fiorica JV [48]	TAM/MA	56	21	5	26

CR complete response, *PR* partial response, *OR* overall response, *MPA* medroxyprogesterone acetate, *TAM* tamoxifen, *MA* megestrole acetate

series and has not shown any clinical advantage compared with either treatment alone [24]. The activity of progestins in this condition is often limited by the frequent down-regulation of PR within the target tissues, resulting in a relatively short duration of response. Tamoxifen can increase PR content in endometrial cancer tissues, but clinical studies on alternating treatment with tamoxifen and progestin have given conflicting results [47–49]. Therapeutic strategies targeted at enhancing PR expression are currently being investigated worldwide and could potentially improve the clinical outcome of endometrial cancer patients (Table 11.6) [50].

4 Novel Approaches

4.1 *Fourth-Generation Progestins*

The fourth-generation progestin dienogest is an effective means of treating endometriosis. At present, its anti-tumor activity is also attracting close attention following a report that dienogest suppresses the proliferation of endometrial cancer-derived cell lines *in vitro* which fail to respond to other progestins such as medroxyprogesterine acetate (MPA) [51]. The mechanism for antitumor activity of dienogest appears different to conventional progestin preparations. It has been shown to suppress neovascularization [52, 53], the cell cycle [54] and to inhibit PGE2 formation through selective antagonist activity on the PR [55].

4.2 *Progesterone Receptor Expression and Reversal of Progesterone Resistance*

Many patients with endometrial cancer are resistant to progestin therapy apparently associated with the absence of the progesterone receptor (PR). Studies aimed at the restoration of PR expression in endometrial cancer have been conducted at the gene and protein levels. Several preclinical studies have been carried out on the control of epigenetic mutations (hypomethylation or hypermethylation) often seen in cases of endometrial cancer for the purpose of stimulating apoptosis and restoration of susceptibility of the cancer to chemotherapy [56]. PR gene hypermethylation, responsible for the disappearance of the PR in certain endometrial cancers, can be reversed by DNA methyltransferase inhibitors, shown to stimulate re-expression of the PR at both the mRNA level and the protein level [57–59]. Epidermal growth factor receptor (EGFR) has also been implicated as a factor involved in progesterone resistance. The EGFR has been detected in histological specimens and cell lines of endometrial cancer and is known to be over expressed in endometrial cancer, although its role in resistance to progestin has not been clarified. One study analyzed differences in EGFR function and resistance to progestin in relation to the presence or absence of PR expression in endometrial cancer, reporting that EGFR was detected in 60 % of PR positive specimens and 90.5 % of PR negative specimens. Furthermore, when further EGFR expression was stimulated in Ishikawa cells (an endometrial cancer cell line), the susceptibility to progestin decreased, accompanied by a reduction in PR expression [60]. AG1478 (a specific inhibitor for EGFR tyrosine kinase) effectively suppressed the proliferation of EGFR over expression in endometrial cancer cells [60]. On the basis of these findings, it is assumed that excessive expression of EGFR in endometrial cancer cells can reduce the susceptibility to progestin therapy. Therefore, inhibitors specific to EGFR tyrosine kinase may be effective against endometrial cancer resistant to progestin therapy.

5 Conclusions

Hysterectomy remains the gold standard for patients with endometrial cancer, but may not be an acceptable option for young women who wish to preserve their fertility or for women with severe co-morbidities compromising surgical survival.

Cohort studies with fertility-sparing treatment followed by assisted reproduction show a high chance of disease regression and encouraging live birth rates for patients with early-stage endometrial cancer. The risk of disease relapse during follow-up is significant and women wanting to pursue this treatment need to undergo thorough counseling.

Progestins are probably not curative, because the underlying cause usually persists. Based on our present understanding of the disease, hysterectomy is advocated once family planning is complete. The disease remains confined to the endometrium at the time of hysterectomy in the overwhelming majority of patients, and the outcome and survival are for the vast majority not jeopardized by conservative treatment [36].

Although hormonal management of complex atypical hyperplasia and low-grade, apparent early-stage endometrial carcinoma has been utilized for 50 years, many questions remain unanswered. Large prospective trials are presently underway to clarify some of the unresolved issues, including the role for levonogestrel containing intrauterine devices (Table 11.5). Similar to other rare cancers, an international registry would further advance our understanding. Investigations on novel therapeutic options targeting the underlying causes and molecular pathways are eagerly awaited.

References

1. Jemal A, et al. Cancer statistics, 2010. CA Cancer J Clin. 2010;60(5):277–300.
2. Duska LR, et al. Endometrial cancer in women 40 years old or younger. Gynecol Oncol. 2001;83(2):388–93.
3. Gitsch G, et al. Endometrial cancer in premenopausal women 45 years and younger. Obstet Gynecol. 1995;85(4):504–8.
4. Lajer H, et al. Survival after stage IA endometrial cancer; can follow-up be altered? A prospective nationwide Danish survey. Acta Obstet Gynecol Scand. 2012;91(8):976–82.
5. Mourits MJ, et al. Safety of laparoscopy versus laparotomy in early-stage endometrial cancer: a randomised trial. Lancet Oncol. 2010;11(8):763–71.
6. Obermair A, et al. Total laparoscopic hysterectomy versus total abdominal hysterectomy for obese women with endometrial cancer. Int J Gynecol Cancer. 2005;15(2):319–24.
7. Cade TJ, et al. Progestogen treatment options for early endometrial cancer. BJOG. 2010;117(7):879–84.
8. Baker J, et al. Efficacy of oral or intrauterine device-delivered progestin in patients with complex endometrial hyperplasia with atypia or early endometrial adenocarcinoma: a meta-analysis and systematic review of the literature. Gynecol Oncol. 2012;125(1):263–70.
9. Matthews TJ, Hamilton BE. Delayed childbearing: more women are having their first child later in life. NCHS Data Brief. 2009;21:1–8.
10. Lee NK, et al. Prognostic factors for uterine cancer in reproductive-aged women. Obstet Gynecol. 2007;109(3):655–62.
11. Walsh C, et al. Coexisting ovarian malignancy in young women with endometrial cancer. Obstet Gynecol. 2005;106(4):693–9.
12. Koskas M, et al. Safety of uterine and/or ovarian preservation in young women with grade 1 intramucous endometrial adenocarcinoma: a comparison of survival according to the extent of surgery. Fertil Steril. 2012;98(5):1229–35.
13. Leitao Jr MM, et al. Comparison of D&C and office endometrial biopsy accuracy in patients with FIGO grade 1 endometrial adenocarcinoma. Gynecol Oncol. 2009;113(1):105–8.
14. Daniel AG, Peters 3rd WA. Accuracy of office and operating room curettage in the grading of endometrial carcinoma. Obstet Gynecol. 1988;71(4):612–4.
15. Sironi S, et al. Myometrial invasion by endometrial carcinoma: assessment by MR imaging. AJR Am J Roentgenol. 1992;158(3):565–9.
16. Wu LM, et al. Predictive value of T2-weighted imaging and contrast-enhanced MR imaging in assessing myometrial invasion in endometrial cancer: a pooled analysis of prospective studies. Eur Radiol. 2013;23(2):435–49.
17. Manfredi R, et al. Local-regional staging of endometrial carcinoma: role of MR imaging in surgical planning. Radiology. 2004;231(2):372–8.
18. Gotlieb WH. Fertility preserving treatments for endometrial cancer: the unanswered questions. Gynecol Oncol. 2013;129(1):1–2.
19. Kim CH, et al. Sentinel lymph node mapping with pathologic ultrastaging: a valuable tool for assessing nodal metastasis in low-grade endometrial cancer with superficial myoinvasion. Gynecol Oncol. 2013;131(3):714–9.
20. Lancaster JM, et al. Society of Gynecologic Oncologists Education Committee statement on risk assessment for inherited gynecologic cancer predispositions. Gynecol Oncol. 2007;107(2):159–62.

21. Kauff ND. How should women with early-onset endometrial cancer be evaluated for lynch syndrome? J Clin Oncol. 2007;25(33):5143–6.
22. Park JY, et al. Long-term oncologic outcomes after fertility-sparing management using oral progestin for young women with endometrial cancer (KGOG 2002). Eur J Cancer. 2013;49(4):868–74.
23. Penner KR, et al. Predictors of resolution of complex atypical hyperplasia or grade 1 endometrial adenocarcinoma in premenopausal women treated with progestin therapy. Gynecol Oncol. 2012;124(3):542–8.
24. Markman M. Hormonal therapy of endometrial cancer. Eur J Cancer. 2005;41(5):673–5.
25. Janzen DM, et al. Progesterone receptor signaling in the microenvironment of endometrial cancer influences its response to hormonal therapy. Cancer Res. 2013;73(15):4697–710.
26. Thigpen JT, et al. Oral medroxyprogesterone acetate in the treatment of advanced or recurrent endometrial carcinoma: a dose-response study by the Gynecologic Oncology Group. J Clin Oncol. 1999;17(6):1736–44.
27. Ramirez PT, et al. Hormonal therapy for the management of grade 1 endometrial adenocarcinoma: a literature review. Gynecol Oncol. 2004;95(1):133–8.
28. Giannopoulos T, Butler-Manuel S, Tailor A. Levonorgestrel-releasing intrauterine system (LNG-IUS) as a therapy for endometrial carcinoma. Gynecol Oncol. 2004;95(3):762–4.
29. Laurelli G, et al. Conservative treatment of early endometrial cancer: preliminary results of a pilot study. Gynecol Oncol. 2011;120(1):43–6.
30. Kim MK, et al. Combined medroxyprogesterone acetate/levonorgestrel-intrauterine system treatment in young women with early-stage endometrial cancer. Am J Obstet Gynecol. 2013;209(4):358e1–4.
31. Minig L, et al. Progestin intrauterine device and GnRH analogue for uterus-sparing treatment of endometrial precancers and well-differentiated early endometrial carcinoma in young women. Ann Oncol. 2011;22(3):643–9.
32. Gunderson CC, et al. Oncologic and reproductive outcomes with progestin therapy in women with endometrial hyperplasia and grade 1 adenocarcinoma: a systematic review. Gynecol Oncol. 2012;125(2):477–82.
33. Ashino-Fuse H, et al. Medroxyprogesterone acetate, an anti-cancer and anti-angiogenic steroid, inhibits the plasminogen activator in bovine endothelial cells. Int J Cancer. 1989;44(5):859–64.
34. Ushijima K, et al. Multicenter phase II study of fertility-sparing treatment with medroxyprogesterone acetate for endometrial carcinoma and atypical hyperplasia in young women. J Clin Oncol. 2007;25(19):2798–803.
35. Kelley RM, Baker WH. Progestational agents in the treatment of carcinoma of the endometrium. N Engl J Med. 1961;264:216–22.
36. Gallos ID, et al. Regression, relapse, and live birth rates with fertility-sparing therapy for endometrial cancer and atypical complex endometrial hyperplasia: a systematic review and meta-analysis. Am J Obstet Gynecol. 2012;207(4):266e1–12.
37. Park JY, et al. Progestin re-treatment in patients with recurrent endometrial adenocarcinoma after successful fertility-sparing management using progestin. Gynecol Oncol. 2013;129(1):7–11.
38. Kim ML, Seong SJ. Clinical applications of levonorgestrel-releasing intrauterine system to gynecologic diseases. Obstet Gynecol Sci. 2013;56(2):67–75.
39. Banno K, et al. Progestin therapy for endometrial cancer: the potential of fourth-generation progestin (review). Int J Oncol. 2012;40(6):1755–62.
40. Varma R, et al. The effectiveness of a levonorgestrel-releasing intrauterine system (LNG-IUS) in the treatment of endometrial hyperplasia–a long-term follow-up study. Eur J Obstet Gynecol Reprod Biol. 2008;139(2):169–75.
41. Montz FJ, et al. Intrauterine progesterone treatment of early endometrial cancer. Am J Obstet Gynecol. 2002;186(4):651–7.
42. Kim MK, et al. Comparison of dilatation & curettage and endometrial aspiration biopsy accuracy in patients treated with high-dose oral progestin plus levonorgestrel intrauterine system for early-stage endometrial cancer. Gynecol Oncol. 2013;130(3):470–3.

43. Wright JD, et al. Safety of ovarian preservation in premenopausal women with endometrial cancer. J Clin Oncol. 2009;27(8):1214–9.
44. Kauppila A. Progestin therapy of endometrial, breast and ovarian carcinoma. A review of clinical observations. Acta Obstet Gynecol Scand. 1984;63(5):441–50.
45. Piver MS, et al. Medroxyprogesterone acetate (Depo-Provera) vs. hydroxyprogesterone caproate (Delalutin) in women with metastatic endometrial adenocarcinoma. Cancer. 1980;45(2):268–72.
46. Podratz KC, et al. Effects of progestational agents in treatment of endometrial carcinoma. Obstet Gynecol. 1985;66(1):106–10.
47. Whitney CW, et al. Phase II study of medroxyprogesterone acetate plus tamoxifen in advanced endometrial carcinoma: a Gynecologic Oncology Group study. Gynecol Oncol. 2004;92(1):4–9.
48. Fiorica JV, et al. Phase II trial of alternating courses of megestrol acetate and tamoxifen in advanced endometrial carcinoma: a Gynecologic Oncology Group study. Gynecol Oncol. 2004;92(1):10–4.
49. Pandya KJ, et al. Megestrol and tamoxifen in patients with advanced endometrial cancer: an Eastern Cooperative Oncology Group Study (E4882). Am J Clin Oncol. 2001;24(1):43–6.
50. Gadducci A, Cosio S, Genazzani AR. Old and new perspectives in the pharmacological treatment of advanced or recurrent endometrial cancer: hormonal therapy, chemotherapy and molecularly targeted therapies. Crit Rev Oncol Hematol. 2006;58(3):242–56.
51. Katsuki Y, et al. Dienogest, a novel synthetic steroid, overcomes hormone-dependent cancer in a different manner than progestins. Cancer. 1997;79(1):169–76.
52. Nakamura M, et al. Dienogest, a synthetic steroid, suppresses both embryonic and tumor-cell-induced angiogenesis. Eur J Pharmacol. 1999;386(1):33–40.
53. Katayama H, et al. Effect of dienogest administration on angiogenesis and hemodynamics in a rat endometrial autograft model. Hum Reprod. 2010;25(11):2851–8.
54. Shimizu Y, et al. Dienogest, a synthetic progestin, inhibits the proliferation of immortalized human endometrial epithelial cells with suppression of cyclin D1 gene expression. Mol Hum Reprod. 2009;15(10):693–701.
55. Shimizu Y, et al. Dienogest, a synthetic progestin, inhibits prostaglandin E2 production and aromatase expression by human endometrial epithelial cells in a spheroid culture system. Steroids. 2011;76(1–2):60–7.
56. Yang S, et al. Endometrial cancer: reviving progesterone therapy in the molecular age. Discov Med. 2011;12(64):205–12.
57. Ren Y, et al. Down-regulation of the progesterone receptor by the methylation of progesterone receptor gene in endometrial cancer cells. Cancer Genet Cytogenet. 2007;175(2):107–16.
58. Xiong Y, et al. Histone deacetylase inhibitors decrease DNA methyltransferase-3B messenger RNA stability and down-regulate de novo DNA methyltransferase activity in human endometrial cells. Cancer Res. 2005;65(7):2684–9.
59. Balch C, et al. Role of epigenomics in ovarian and endometrial cancers. Epigenomics. 2010;2(3):419–47.
60. Ai Z, et al. Overexpressed epidermal growth factor receptor (EGFR)-induced progestin insensitivity in human endometrial carcinoma cells by the EGFR/mitogen-activated protein kinase signaling pathway. Cancer. 2010;116(15):3603–13.
61. Bokhman JV, et al. Can primary endometrial carcinoma stage I be cured without surgery and radiation therapy? Gynecol Oncol. 1985;20(2):139–55.
62. Randall TC, Kurman RJ. Progestin treatment of atypical hyperplasia and well-differentiated carcinoma of the endometrium in women under age 40. Obstet Gynecol. 1997;90(3):434–40.
63. Kim YB, et al. Progestin alone as primary treatment of endometrial carcinoma in premenopausal women. Report of seven cases and review of the literature. Cancer. 1997;79(2):320–7.
64. Imai M, et al. Medroxyprogesterone acetate therapy for patients with adenocarcinoma of the endometrium who wish to preserve the uterus-usefulness and limitations. Eur J Gynaecol Oncol. 2001;22(3):217–20.
65. Kaku T, et al. Conservative therapy for adenocarcinoma and atypical endometrial hyperplasia of the endometrium in young women: central pathologic review and treatment outcome. Cancer Lett. 2001;167(1):39–48.

66. Wang CB, et al. Fertility-preserving treatment in young patients with endometrial adenocarcinoma. Cancer. 2002;94(8):2192–8.
67. Gotlieb WH, et al. Outcome of fertility-sparing treatment with progestins in young patients with endometrial cancer. Obstet Gynecol. 2003;102(4):718–25.
68. Jadoul P, Donnez J. Conservative treatment may be beneficial for young women with atypical endometrial hyperplasia or endometrial adenocarcinoma. Fertil Steril. 2003;80(6):1315–24.
69. Niwa K, et al. Outcome of fertility-preserving treatment in young women with endometrial carcinomas. BJOG. 2005;112(3):317–20.
70. Ota T, et al. Clinicopathologic study of uterine endometrial carcinoma in young women aged 40 years and younger. Int J Gynecol Cancer. 2005;15(4):657–62.
71. Yahata T, et al. Long-term conservative therapy for endometrial adenocarcinoma in young women. Hum Reprod. 2006;21(4):1070–5.
72. Yang YC, et al. Reevaluating the safety of fertility-sparing hormonal therapy for early endometrial cancer. Gynecol Oncol. 2005;99(2):287–93.
73. Le Digabel JF, et al. Young women with atypical endometrial hyperplasia or endometrial adenocarcinoma stage I: will conservative treatment allow pregnancy? Results of a French multicentric survey. Gynecol Obstet Fertil. 2006;34(1):27–33.
74. Elizur SE, et al. Outcome of in vitro fertilization treatment in infertile women conservatively treated for endometrial adenocarcinoma. Fertil Steril. 2007;88(6):1562–7.
75. Minaguchi T, et al. Combined phospho-Akt and PTEN expressions associated with post-treatment hysterectomy after conservative progestin therapy in complex atypical hyperplasia and stage Ia, G1 adenocarcinoma of the endometrium. Cancer Lett. 2007;248(1):112–22.
76. Wheeler DT, Bristow RE, Kurman RJ. Histologic alterations in endometrial hyperplasia and well-differentiated carcinoma treated with progestins. Am J Surg Pathol. 2007;31(7):988–98.
77. Yamazawa K, et al. Fertility-preserving treatment with progestin, and pathological criteria to predict responses, in young women with endometrial cancer. Hum Reprod. 2007;22(7):1953–8.
78. Li HZ, Chen XN, Qiao J. Letrozole as primary therapy for endometrial hyperplasia in young women. Int J Gynaecol Obstet. 2008;100(1):10–2.
79. Eftekhar Z, et al. Efficacy of megestrol acetate (megace) in the treatment of patients with early endometrial adenocarcinoma: our experiences with 21 patients. Int J Gynecol Cancer. 2009;19(2):249–52.
80. Hahn HS, et al. Conservative treatment with progestin and pregnancy outcomes in endometrial cancer. Int J Gynecol Cancer. 2009;19(6):1068–73.
81. Han AR, et al. Pregnancy outcomes using assisted reproductive technology after fertility-preserving therapy in patients with endometrial adenocarcinoma or atypical complex hyperplasia. Int J Gynecol Cancer. 2009;19(1):147–51.
82. Signorelli M, et al. Fertility-sparing treatment in young women with endometrial cancer or atypical complex hyperplasia: a prospective single-institution experience of 21 cases. BJOG. 2009;116(1):114–8.
83. Yu M, et al. Outcome analysis of conservative treatment of well-differentiated endometrial adenocarcinoma and severe atypical hyperplasia in young women. Zhonghua Fu Chan Ke Za Zhi. 2006;41(4):242–5.
84. Mao Y, et al. Outcomes of conservative therapy for young women with early endometrial adenocarcinoma. Fertil Steril. 2010;93(1):283–5.
85. Mazzon I, et al. Conservative surgical management of stage IA endometrial carcinoma for fertility preservation. Fertil Steril. 2010;93(4):1286–9.
86. Park H, et al. Effectiveness of high-dose progestin and long-term outcomes in young women with early-stage, well-differentiated endometrioid adenocarcinoma of uterine endometrium. Arch Gynecol Obstet. 2012;285(2):473–8.
87. Perri T, et al. Prolonged conservative treatment of endometrial cancer patients: more than 1 pregnancy can be achieved. Int J Gynecol Cancer. 2011;21(1):72–8.
88. Thigpen T, et al. Tamoxifen in the treatment of advanced or recurrent endometrial carcinoma: a Gynecologic Oncology Group study. J Clin Oncol. 2001;19(2):364–7.

Chapter 12
Progestogens and the Menopause

Eitan Pe'er

1 Introduction

Progesterone is an essential hormone during the reproductive years. It does not have any physiological role during menopause. Yet many synthetic progestogens, are in clinical use during the menopause and post menopause, most notably as adjunct to estrogen replacement therapy. Progestogens are necessary in order to prevent endometrial hyperplasia caused by unopposed estrogen [1]. In recent years, large clinical trials have shown that the association of progestins with estrogen in HRT might raise the risk of breast cancer [2, 3], but this risk was not confirmed in ongoing clinical studies which have not shown that trend [4]. Newer evidence from molecular and genomic studies suggests that not all progestins have the same effect. As shown in other chapters in this book, there is no "class-effect" of progestogens. Each progestin has its own biocharcteristics regarding breast cancer as in other clinical conditions.

In recent years clinical and new basic data have clearly shown that progesterone has many other properties and functions beside reproduction and pregnancy. Progesterone has been shown to have neuroprotective effects [5] and is used in various brain injuries [6].

Progesterone has both genomic and non-genomic actions. The genomic action is the classical reaction whereby the hormone progesterone, acting as a ligand, connects to its receptor in the nucleus and initiates new mRNA protein synthesis. This is a relatively slow process. The rapid action of progestogens are due to non-genomic

E. Pe'er, M.D. (✉)
Gyneco-Endocrinology Unit, Rambam Health Campus,
Technion School of Medicine, Haifa, Israel
e-mail: eitanpeer@gmail.com

H.J.A. Carp (ed.), *Progestogens in Obstetrics and Gynecology*,
DOI 10.1007/978-3-319-14385-9_12

actions in which intracellular signaling pathways are activated resulting in alteration of ions fluxes and intracellular calcium concentration within seconds [7]. The non genomic actions also induce second messengers, such as cyclic nucleotides and extracellular regulated kinases [8]. These actions are specific to each individual progestogen. As progestogens differ in actions, proper selection is necessary for successful therapy in treating menopausal symptoms or reducing the risk of diseases associated with the menopause.

2 Progestogens and Osteoporosis

Progestogens alone have a very limited effect on bone mineral density (BMD). But, pretreatment with estrogen for 4–7 days has been shown to induce progesterone receptors in osteoblasts [9]. This explains the beneficial effect on BMD, in all clinical trials in which progestogens were added to estrogen.

Medroxyprogesterone acetate (MPA) decreases BMD, both in the hips and spine, in the first 2 years of use, followed by a slight increase [10]. The effect of norethisterone acetate (NETA) with a low dose of 17β-estardiol on BMD has been investigated in a randomized placebo control study [11]. There was a significant increase in BMD both in the lumbar spine (5.2 %) and the hip (3.1 %) compared to the placebo group (−0.9%). Serum concentration of osteocalcin decreased by approximately 34 %, bone-specific alkaline phosphatase decreased by about 30 %, and C-terminal propepetide of type I collagen decreased by 20 %. Norethisterone is a synthetic progestin derived from 19-nortestosterone. It binds and activates the progesterone receptor twice as much as progesterone itself, with low androgenic and estrogenic activities attributed to its metabolites [12]. NETA alone has been studied in male castrated mice. In this study NETA alone was found to have a slightly protective effect against bone loss [13]. Controversially, inhibition of the nuclear progesterone receptor has been found to augment bone mass, resulting in higher BMD [14].

3 Progesterone, Hot Flush and Night Sweats

Nightly micronized progesterone (300 mg) has been shown to cause an overall decrease in the number of daily hot flushes and night sweats by 59 % [15]. Extension of this study has shown that micronized progesterone is also effective for severe vasomotor symptoms and that progesterone withdrawal is not followed by a rebound increase in vasomotor symptoms [16]. A similar effect has been shown for medroxyprogesterone acetate (MPA) 10 mg/day. Prior et al. [17], carried out a randomized double-blind trial in women after overiectomy. Patients received either conjugated equine estrogen (CEE), (0.6 mg/day) or MPA for 1 year. MPA was found to be equivalent to CEE in the control of vasomotor symptoms in women treated immediately following the surgery [17]. Oral progestins have been shown to be effective for vasomotor symptoms in several randomized placebo controlled trial [18, 19].

Estrogen alleviates hot flushes by lowering levels of serotonine and noradrenaline in the brain [19]. Progestogens may act in a different manner. Progesterone acts on the hypothalamus changing the frequency of LH pulses, increasing basal temperature and stimulating respiration. Therefore progesterone and progestins have various effects on the hypothalamus, from which hot flushes are thought to generate [20].

4 Progesterone and Venous Thromboembolism

Progestins, when given alone (e.g. progestin only contraceptive pills), carry little, if any, risk of VTE. Randomized controlled studies and meta-analyses of observational studies suggest that the risk of venous thrombo-embolism (VTE) is higher among users of combined estrogen and progestogen than among users of estrogen alone. Oral estrogens increase the VTE risk while transdermal estrogens appear to be safe with respect to thrombotic risk [21]. However, MPA, by activating glucocorticoid receptors, potentiate the vascular effects of thrombin [22]. The ESTHER study (Estrogen and Thromboembolism Risk) looked into the risk of VTE in French postmenopausal women treated with HRT. This study was the first to establish a differential association of VTE risk related to the progestogen used. The results were irrespective of the route of estrogen administration. The results showed that micronized progesterone and pregnane derivates are safer with regard to VTE risk. The norpregnane derivatives were associated with a significant increase in VTE risk [23].

The norpregnanes are potent progestogens with antiestrogenic activity. Women suffering from hyperestrogenic effects, such as breast tenderness or endometrial hyperplasia are more likely to benefit from this type of progestogens.

5 Progesterone and the Brain

Progesterone is metabolized in the brain by 5α-reductase to 5α-dihydroprogesterone. This in turn is further metabolized by 3α-hydroxysteroid dehydrogenase to the neurosteroid allopregnanolone.

The neurosteroids are modulatory ligands for a variety of neurotransmitters and nuclear steroid hormone receptors. Allopregnanolone crosses the blood-brain barrier. It has been shown in rodents, that allopregnanolone is an efficacious proliferative agent (both in vitro and in-vivo studies) [24]. It also decreases amyloid protein in human neural stem cells [25]. Progesterone metabolites exert considerable sedative effects after binding to the GABAA receptor [26]. Neurosteroids such as pregnanolone affect synaptic functions and myelinization. Their action is mediated through inhibition of the glycogen synthase kinase (GSK-3β) pathway (much like most bipolar mood stabilizers such as lithium) [27]. Neurosteroids are modulatory ligands for a variety of neurotransmitters and nuclear steroid hormone receptors. In a mice model with experimental autoimmune encephalomyelitis it has been shown that both the spinal cord and the brain are sensitive to the protective effects of progesterone. Progesterone has been shown to reduce inflammatory reactions commonly seen in MS, by the direct effect of progesterone on astrocytes and microglia [28].

5.1 *Progesterone in Alzheimer's Disease*

Estradiol increases the expression of the progesterone-synthesizing enzymes. Estradiol increases the expression in the hypothalamus, and especially in the astrocytes. Astrocytes are the most active steroidogenic cells in CNS and contribute to neuro-protection [29]. Treatment with different types of progestogens found that these compounds may promote neurogenesis, neural survival, myelinization and increases memory [30]. There is some data that suggests that allopregnanolone may maintain the regenerative ability of the brain and also can modify the progression of Alzheimer's disease [31]. Progesterone has been shown to improve impaired axonal transport, a key event of the aging brain. Reduced axonal transport has been proposed to play an early and causative role in the development of Alzheimer's disease. In mouse models, reduced axonal transport may lead to aberrant amyloid-β peptide formation and subsequently to neurodegeneration [32].

5.2 *Traumatic Brain Injury (TBI)*

In TBI, whether post menopausal or not, the use of progesterone was found to be effective in reducing brain damage. This subject is fully discussed in Chap. 14. In brief, data is available showing that progesterone reduces edema, restores the blood-brain barrier, protects against secondary neuronal death and promotes behavioral recovery after TBI [33, 34]. A phase II clinical trial, the ProTECT study, has shown more than 50 % reduction in mortality in mortality in severe TBI and a statistically significant improvement in functional outcome in patients with moderate BTI, when treatment was administered no later than 2 h after sustaining BTI [35].

6 Conclusion

Progesterone and progestogens have a significant role in various clinical situations throughout life. As the novel actions of progestogens are elucidated it is clear that these hormones have influence on the outcome of many clinical conditions.

It is important to bear in mind that there is no class effect of all these compounds. Each has its own clinical, biochemical and molecular specific effects. In depth knowledge of the physio-pathological effects of each progestogen will enable their better use in many clinical conditions such as hot flushes, brain trauma, sleep disorder and more. The most serious clinical side effects are the raised risk of breast cancer associated with some progestogens (e.g. MPA) but not with others such as micronized progesterone. MPA has also been shown to inhibit some of the beneficial effects of estradiol on the CNS. The fact that numerous coregulators affect the end result of ligand-progesterone receptors (both nuclear and membrane) indicates to the complexities of progesterone/progestogens actions. Receptor affinity alone does not determine potency. Our present understanding is that the affinity, potency and efficacy of progestogens are substantially different between the different types of progestogens and are tissue specific.

References

1. Ziel HK, Finkle WD. Association of estrone with the development of endometrial carcinoma. Am J Obstet Gynecol. 1976;124(7):735–40.
2. Rossouw JE et al. Risks and benefits of estrogen plus progestin in healthy postmenopausal women. Principal results from the Women's Health Initiative randomized controlled study. JAMA. 2002;288:321–33.
3. Beral V. Breast cancer and hormone-replacement therapy in the Million Women Study. Lancet. 2003;362:419–27.
4. Fournier A et al. Breast cancer risk in relation to different types of hormone replacement therapy in the E3N-EPIC cohort. Int J Cancer. 2005;114:448–54.
5. Feeser VR, Loria RM. Modulation of traumatic brain injury using progesterone and the role of glial cells on its neuroprotective actions. J Neuroimmunol. 2011;237:4–12.
6. Stein DG, Wright DW. Progesterone in the clinical treatment of acute traumatic brain injury. Expert Opin Investig Drugs. 2010;19:847–57.
7. Blackmore PF et al. Cell surface-binding sites for progesterone mediated calcium-uptake in human sperm. J Biol Chem. 1991;266:18655–9.
8. Zhu Y et al. Cloning, expression, and characterization of a membrane progestin receptor and evidence it is an intermediary in meiotic maturation of fish oocytes. Proc Natl Acad Sci U S A. 2003;100:2231–6.
9. Prior FC. Progesterone as a bone-trophic hormone. Endocr Rev. 1990;11:386–98.
10. Clark MK et al. Bone mineral density loss and recovery during 48 months in first-time users of depo medroxyprogesterone acetate. Fertil Steril. 2006;86:1466–74.
11. Delmas PD et al. A combination of low doses of 17 beta-estradiol and northisterone acetate prevents bone loss and normalizes bone turnover in postmenopausal women. Osteoporos Int. 2000;11:177–87.
12. Schoonen WG et al. Hormonal properties of norethisterone, 7alpha-methyl-norethisterone and their derivates. J Steroid Biochem Mol Biol. 2000;74:213–22.
13. Broulik PD et al. Progestagens androgenic action on the bone of male castrated mice. Prague Med Rep. 2006;107:401–8.
14. Yao W et al. Inhibition of nuclear receptor during the bone linear growth phase increases peak bone mass in female mice. PLoS One. 2010;5:e11410.
15. Hitchcock CL, Prior JC. Oral micronized progesterone for vasomotor symptoms. A placebo-controlled randomized trial in healthy postmenopausal women. Menopause. 2012;10(1097).

16. Prior JC, Hitchcock CL. Progesterone for hot flush and night sweat treatment- effectiveness for severe vasomotor symptoms and lack of withdrawal rebound. Gynecol Endocrinol. 2012;28(s2):7–11.
17. Prior JC, Nielsen JD, et al. Medroxyprogesterone and conjugated oestrogen are equivalent for hot flushes: a 1-year randomized Double-blind trial following premenopausal overiectomy. Clin Sci (Lond). 2007;112:517–25.
18. Loprinzi CL et al. Megesterol acetate for the prevention of hot flashes. N Engl J Med. 1994;331:347–52.
19. Dennerstein L et al. Menopausal hot flashes: a double blind comparison of placebo, ethinyl estardiol and norgestrel. Br J Obstet Gynaecol. 1978;85:852–6.
20. Freedman RR. Biochemical, metabolic, and vascular mechanisms in menopausal hot flushes. Fertil Steril. 1998;70:332–7.
21. Olie V et al. Risk of venous thrombosis with oral versus transdermal estrogen therapy among postmenopausal women. Curr Opin Hematol. 2010;17:457–63.
22. Stanczyk FZ et al. Progestogens used in postmenopausal hormone therapy: differences in their pharmacological properties, intracellular actions and clinical effects. Endocr Rev. 2013; 34:171–208.
23. Canonico M et al. Hormone therapy and venous thromboembolism among postmenopausal women: impact of the route of estrogen administration and progestogens: the ESTHER study. Circulation. 2007;115:840–5.
24. Chen S et al. Allopregnanolone promotes regeneration and reduces β-amyloid burden in a preclinical model of Alzheimer's disease. PLoS One. 2011;6:e24293.
25. Brinton RD, Wang JM. Therapeutic potential of neurogenesis for prevention and recovery from Alzheimer's disease: allopregnanolone as a proof of concept neurogenic agent. Curr Alzheimer Res. 2006;3:185.
26. Kuhl H. Pharmacology of estrogens and progestogens: influence of different routes of administration. Climacteric. 2005;8(suppl):3–63.
27. Chuang DM et al. GSK-3 as a target for lithium induced neuroprotection against excitotoxicity in neural cultures and animal models of ischemic stroke. Front Mol Neurosci. 2011;4:15T.
28. Garay LI et al. Progesterone down-regulates spinal cord inflammatory mediators and increases myelination in experimental autoimmune encephalomyelitis. Neuroscience. 2012;226:40–50.
29. Compagnone NA, Mellon SH. Neorosteroids: biosynthesis - and function of these novel neoromodulators. Front Neuroendocrinol. 2000;21:1–58.
30. Mellon SH. Neurosteroid regulation of CNS development. Pharmacol Ther. 2007;116: 107–24.
31. Micevych P, Soma KK, Sinchak K. Neuroprogesterone; key to estrogen positive feedback? Brain Res Rev. 2008;57(2):470–80.
32. Wang JM et al. Regenerative potential of allopregnanolone. Brain Res Rev. 2008; 57:398–409.
33. Herson PS et al. Sex, sex steroids and brain injury. Semin Reprod Med. 2009;27:229–39.
34. Roof RL et al. Progesterone facilitates cognitive recovery and reduces secondary neuronal loss caused by cortical contusion injury in male rats. Exp Neurol. 1994;129:64–9.
35. Wright DW et al. ProTECT: a randomized clinical trial of progesterone for acute traumatic brain injury. Ann Emerg Med. 2007;49:391–402.

Chapter 13
Progestogens and Autoimmunity

Abraham Tsur, Grant C. Hughes, and Yehuda Shoenfeld

1 The Impact of Sex Hormones on the Immune System

The human immune system exhibits sexual dimorphism. Approximately 78 % of patients affected with an autoimmune disease (AD) are women [1]. The female to male ratio may vary from 9:1 and higher in systemic lupus erythematous (SLE), Sjögren's syndrome and autoimmune thyroid disease, to 3:1 in multiple sclerosis (MS) and rheumatoid arthritis (RA). Only a few ADs are more common in men, but include the spondyloarthropathies, autoimmune diabetes and Guillain-Barré syndrome [2, 3]. Moreover, men and women show different susceptibilities to allergy and infection, and these differences are influenced by hormonal status [4, 5].

The immunomodulatory effects of sex hormones are a major factor leading to the sexual dimorphism described above. Non-hormonal factors are also important. These include sex-dependent expression of genes encoded on X [6] and Y [7] chromosomes, microchimerism [8] and gender related behavioral factors. The hypothesis that sex hormones are a major player associated with sexual immune dimorphism is supported by the observation that many ADs appear, fluctuate or resolve when there are hormonal changes, such as at the beginning of the menstrual cycle, during the cycle, pregnancy or puerperium, and following hormone replacement therapy (HRT) or hormonal contraception [3]. Moreover, the risk of certain allergic diseases

A. Tsur, M.D. (✉)
Obstetrics and Gynecology, Sheba Medical Center, Tel Hshomer, Israel
e-mail: avitsur@gmail.com

G.C. Hughes, M.D.
Division of Rheumatology, Department of Medicine, University of Washington, Seattle, WA, USA

Y. Shoenfeld, M.D., F.R.C.P.
Zabludowicz Center for Autoimmune Diseases, Sheba Medical Center, Tel-Hashomer, Israel

Sackler School of Medicine, Tel Aviv University, Tel Aviv, Israel

H.J.A. Carp (ed.), *Progestogens in Obstetrics and Gynecology*,
DOI 10.1007/978-3-319-14385-9_13

and infections is closely associated with hormonal status. At present, there is abundant data regarding the immunomodulatory effect of estrogen, while the data and understanding of the effects of progestogens is lagging behind considerably.

1.1 Progestogens Effects on Different Components of the Immune System

Progestogens, whether natural or synthetic, modulate inflammation, immunity and autoimmunity through direct actions in cells of both the innate and adaptive immune systems. Leukocytes express a broad range of specific hormone receptors, including nuclear and membrane progesterone receptors [9].

Two important targets of progestogen immunomodulation are T cells and B cells. T cells protect against infection by causing the elimination of infected cells and by providing maturation signals to B cells. B cells in turn produce antibodies (Abs), which can neutralize and kill pathogens. In AD, healthy tissues are attacked by T cells and self-reactive Abs, called auto-antibodies (autoAbs). A subset of T cells, T helper (Th) cells, is strongly regulated by progestogens. During pregnancy, high levels of progesterone and estrogens appear to suppress the development and functions of Th1 and Th17 cells while facilitating the development and functions of Th2 cells. Th1 cells are important for defense against intracellular pathogens, but one of their effector molecules, interferon-gamma, appears to be particularly dangerous to the developing fetus [10, 11]. Th17 cells, characterized by production of pro-inflammatory IL-17, are believed to be involved in the chronic inflammation of RA and MS [12], and their activation is associated with recurrent pregnancy loss [13]. During pregnancy, progesterone appears to foster the induction of regulatory Th cells (Tregs) [9], which might prevent harmful responses to the fetus by the maternal immune system [14]. Thus, during pregnancy, progesterone directs the maternal immune response toward increased Treg/Th2 activity and reduced Th1/Th17 activity. While this shift possibly is involved in the remission of rheumatoid arthritis and multiple sclerosis during pregnancy, it could also explain why pregnant women have an increased risk of infection with some intracellular pathogens such as *Listeria monocytogenes*, HSV-2 and HIV [15–17].

Progesterone and estrogen have important effects on B cell differentiation and effector functions. Progesterone suppresses immunoglobulin class switch recombination and somatic hypermutation in B cells. These two processes are required for B cells to make highly protective Abs or highly pathogenic autoAbs. This may be one mechanism whereby progesterone protects against certain autoAb-mediated ADs. Interestingly, estrogen enhances these same pathways, suggesting that the balance between progesterone and estrogen is an important determinant of outcomes in Ab-mediated responses [18]. During pregnancy, progesterone and other pregnancy hormones might induce remission of rheumatoid arthritis by altering post-translational glycosylation of autoAbs, rendering them less capable of inducing inflammation [19, 20].

Progestogens also regulate inflammation through their effects on components of the innate immune system. Upon sensing infection and other triggers, cells of the innate immune system (e.g., dendritic cells and macrophages) release inflammatory mediators and prime the adaptive (memory) response by presenting antigens to immature T cells and B cells. Progesterone appears to program dendritic cells that favor the differentiation of Tregs, but not Th1 or Th17 cells [21]. Importantly, progesterone suppresses the production of several pro-inflammatory cytokines (e.g., IL-1β, TNF-α, IL-6 and IL-23) involved in the pathogenesis of RA by dendritic cells and macrophages. This progesterone action, in concert with up-regulation of endogenous inhibitors of IL-1β and TNF-α, could be an important mechanism of pregnancy-induced RA remission [22]. In MS, progesterone also appears to protect against disease through direct actions on the central nervous system [23]. However, the anti-inflammatory effects of progesterone may also increase the risk for certain infections: Suppression of the innate anti-viral cytokine IFN-alpha by high-level progestogens may explain why pregnant women or women using medroxyprogesterone birth control have a significantly increased risk of acquiring HIV [24].

In summary, Progestogens act on several key components of the immune system to control inflammation, immunity and autoimmunity. During pregnancy, the immunomodulatory actions of progestogens contribute to maternal tolerance of the immunologically semi-allogeneic fetus. Some of these actions may help explain two immunological phenomena associated with pregnancy: the remission of rheumatoid arthritis and multiple sclerosis and increased susceptibility to infection with certain pathogens including HIV.

1.2 Progestogens Effect on Specific Autoimmune Diseases

Progesterone levels vary throughout the different life phases. An initial rise occurs post puberty with the beginning of ovulation and the menstrual cycle. Progesterone levels fluctuate with each subsequent cycle, peaking in the mid-luteal/secretory phase then returning to basal levels during the estrogen-dominant follicular/proliferative phase. The most prominent and prolonged rise in progesterone (and estrogen) levels occurs during pregnancy. After menopause, progesterone levels decline back to their pre-puberty levels. These physiological changes represent an opportunity to examine the effects of progesterone levels on immunity and autoimmune diseases.

Pregnancy is of special interest. The maternal immune system must tolerate the paternal antigens presented by the semi-allogeneic fetus, while maintaining adequate immune defenses to protect both mother and fetus from invading pathogens [25]. Nevertheless the rise of progesterone during pregnancy is accompanied by a rise in the level of estrogens. The only period in which there is a surge of progesterone without a similar surge of estrogen is during the mid luteal phase and this is a limited period of a few days. Therefore examining the impact of progestogens on specific diseases requires a comprehensive view of the different phases and while comparing observations with the mechanism of action of progesterone as described above.

1.2.1 Rheumatoid Arthritis (RA)

Several studies have shown that pregnancy leads to improvement of RA in half to three quarters of patients, followed in many cases with a post partum flare [26, 27]. In addition, it has been demonstrated that during the luteal phase of the ovulatory cycle there is a subjective improvement in morning stiffness and pain accompanying RA [28]. As mentioned above, high systemic progesterone levels during pregnancy may contribute to remission of RA via enhancement of Th2 and Treg activity, decrease of Th1 and Th17 production, and inhibition of the inflammatory cytokines IL-1β, IL-6, TNF-α and IL-23. Interestingly it has been demonstrated that the described improvement occurred more frequently (75 %) in women negative for both ACPA and RF than in women with these antibodies (39 %) [29].

1.2.2 Multiple Sclerosis (MS)

During the second and third trimesters there is a decrease in frequency of MS relapses, followed by an increase in the relapse rate up to 6 months postpartum. The effect of the menstrual cycle on MS is not consistent [30]. As with RA, progesterone may contribute to pregnancy remission of MS through suppression of Th1 and Th17 activity. In addition, progestogens may protect against neuronal damage via non-immunologic mechanisms. Accordingly, a clinical trial is presently being performed examining the potential benefit of progestogens for MS treatment and prevention of relapses [31].

1.2.3 Systemic Lupus Erythematous (SLE)

Unlike RA, SLE does not tend to remit during pregnancy. It is unclear whether pregnancy induces lupus flares: different studies show different conclusions, some showing that pregnant women are at an increased risk for lupus flares while others have reported that the flare rate is unchanged during pregnancy [32]. However, progestogens and estrogens appear to impact on the risk of developing SLE. There is good evidence from large observational studies that exogenous estrogens increase the risk of SLE, possibly by modulating survival and differentiation of self-reactive B cells and T cells [3]. In contrast, there is emerging evidence from animal models that progestogens may protect against the development of SLE by counteracting estrogen's effects on B cells and T cells [9]. Chabbert-Buffet et al. [33] have shown that when the effect of progestogens is isolated from the effect of estrogen (in the form of progestin only contraception), there was a reduced rate of SLE flares. Hence, it seems that estrogens promote SLE while progestogens may have a protective effects.

1.2.4 Autoimmune Thyroid Disease (AITD)

Both Hashimoto's thyroiditis (HT) and Graves's disease also show a very high preponderance in females, which like SLE, is attributed in part to sex hormones. The data regarding AITD and sex hormone is scarce relative to other ADs. In a mouse model of AITD [34] estrogen had a protective effect while progesterone appeared to augment the levels of autoimmunity. In a cross-sectional study among healthy female relatives of patients with autoimmune thyroid disease, estrogen use was associated with a lower risk of AITD [35]. In summary, it appears that compared to SLE, female sex steroids have different effects on the risk of AITD: progestogens may promote AITD while estrogens may have a protective effect in some patients.

2 Progestagen Hypersensitivity

Another aspect of the interaction of progestogens with the immune system is progestagen hypersensitivity, which may present as autoimmune progesterone dermatitis or cyclic urticaria. This rare condition is characterized by a hypersensitive reaction to endogenous or exogenous progesterone/progestins [36–39].

The clinical presentation of progestogen hypersensitivity is mainly dermal but may also by systemic. Table 13.1 summarizes the different manifestations [37, 38].

Exacerbations occur whenever endogenous or exogenous levels of progestogens rise as detailed in Table 13.2. The most classic manifestation is cyclic—appearing at the end of luteal phase of the ovulatory cycle when progesterone levels are high, resolving a few days after menses [37, 38].

The condition is clinically suspected from the cyclic response and/or exacerbation due to external progestins. Confirmations of the diagnosis can be achieved using progesterone skin tests [40, 41].

The classic first line treatment is inhibition of endogenous progesterone secretion by suppression of ovulation. This can be achieved pharmacologically using estrogens, or continuous GNRH agonists. Administration of unopposed estrogens may increase the risk of endometrial carcinoma—thus limiting the use of estrogens [40]. An alternative first line treatment is desensitization with small doses of progesterone

Table 13.1 Clinical manifestations of progesterone hypersensitivity

Dermal manifestations	Systemic manifestations
Erythema multiforme	Progesterone induced anaphylaxis
Eczema	Premenstrual syndrome
Urticaria	Dysmenorrhea
Pruritus	Mastalgia
Angioedema	Headache
Dermatitis	Arthralgia
Acne	Asthma/rhinitis

Table 13.2 Timing of exacerbations of progestogens hypersensitivity

Due to endogenous progesterones rise
Luteal phase of menstrual cycle when progesterone levels are high
Pregnancy (may lead to both improvement or exacerbation)
Due to exogenous rise in progestins
Contraceptive pills
Postmenopausal Hormone replacement therapy

[36, 42]. This approach is appropriate for non dermal manifestations such as dysmenorrhea and premenstrual syndrome [36, 43].

The use of high dose systemic steroids is controversial both because inconsistent data regarding benefit [40] and possible side effects.

In a few patients with refractory symptoms, bilateral oophorectomy has been used. This option may succeed in controlling hypersensitivity symptoms but should be considered to be treatment of last resort [40, 44].

References

1. Fairweather DS, Frisancho-Kiss S, Rose NR. Sex differences in autoimmune disease from a pathological perspective. Am J Pathol. 2008;173:600–9.
2. McCombe PA, Greer JM, Mackay IR. Sexual dimorphism in autoimmune disease. Curr Mol Med. 2009;9:1058–79.
3. Quintero OL, Amador-Patarroyo MJ, Montoya-Ortiz G, Rojas-Villarraga A, Anaya JM. Autoimmune disease and gender: plausible mechanisms for the female predominance of autoimmunity. J Autoimmun. 2012;38:J109–19.
4. Bouman A, Heineman MJ, Faas MM. Sex hormones and the immune response in humans. Hum Reprod Update. 2005;11:411–23.
5. Chen W, Mempel M, Schober W, Behrendt H, Ring J. Gender difference, sex hormones, and immediate type hypersensitivity reactions. Allergy. 2008;63:1418–27.
6. Scofield RH. Genetics of systemic lupus erythematosus and Sjögren's syndrome. Curr Opin Rheumatol. 2009;21:448–53.
7. Bellott DW, Hughes JF, Skaletsky H, Brown LG, Pyntikova T, Cho TJ, et al. Mammalian Y chromosomes retain widely expressed dosage-sensitive regulators. Nature. 2014;508:494–9.
8. Adams KM, Nelson JL. Microchimerism: an investigative frontier in autoimmunity and transplantation. JAMA. 2004;291:1127–31.
9. Hughes GC. Progesterone and autoimmune disease. Autoimmun Rev. 2012;11:A502–14.
10. Gargano J, Holzman C, Senagore P, Thorsen P, Skogstrand K, Hougaard D, Rahbar M, Chung H. Mid-pregnancy circulating cytokine levels, histologic chorioamnionitis and spontaneous preterm birth. J Reprod Immunol. 2008;79:100–10.
11. Moura E, Mattar R, deSouza E, Torloni M, Goncalves-Primo A, Daher S. Inflammatory cytokine gene polymorphisms and spontaneous preterm birth. J Reprod Immunol. 2009;80:41–8.
12. Miossec P, Kolls JK. Targeting IL-17 and TH17 cells in chronic inflammation. Nat Rev Drug Discov. 2012;11:763–76.
13. Lee SK, Kim JY, Lee M, Gilman-Sachs A, Kwak-Kim J. Th17 and regulatory T cells in women with recurrent pregnancy loss. Am J Reprod Immunol. 2012;67:311–8.
14. Arck PC, Hecher K. Fetomaternal immune cross-talk and its consequences for maternal and offspring's health. Nat Med. 2013;19:548–56.

15. Kaushic C, Roth K, Anipindi V, Xiu F. Increased prevalence of sexually transmitted viral infections in women: the role of female sex hormones in regulating susceptibility and immune responses. J Reprod Immunol. 2011;88:205–9.
16. Kourtis AP, Read JS, Jamieson DJ. Pregnancy and infection. N Engl J Med. 2014;370:2211–8.
17. Rowe JH, Ertelt JM, Aguilera MN, Farrar MA, Way SS. Foxp3(+) regulatory T cell expansion required for sustaining pregnancy compromises host defense against prenatal bacterial pathogens. Cell Host Microbe. 2011;10:54–64.
18. Sakiani S, Olsen NJ, Kovacs WJ. Gonadal steroids and humoral immunity. Nat Rev Endocrinol. 2013;9:56–62.
19. Arnold JN, Wormald MR, Sim RB, Rudd PM, Dwek RA. The impact of glycosylation on the biological function and structure of human immunoglobulins. Annu Rev Immunol. 2007;25:21–50.
20. Prados MB, La Blunda J, Szekeres-Bartho J, Caramelo J, Miranda S. Progesterone induces a switch in oligosaccharyltransferase isoform expression: consequences on IgG N-glycosylation. Immunol Lett. 2011;137:28–37.
21. Hughes GC, Clark EA. Regulation of dendritic cells by female sex steroids: relevance to immunity and autoimmunity. Autoimmunity. 2007;40:470–81.
22. Østensen M, Villiger PM. The remission of rheumatoid arthritis during pregnancy. Semin Immunopathol. 2007;29:185–91.
23. Hoffman GE, Merchenthaler I, Zup SL. Neuroprotection by ovarian hormones in animal models of neurological disease. Endocrine. 2006;29:217–31.
24. Hel Z, Stringer E, Mestecky J. Sex steroid hormones, hormonal contraception, and the immunobiology of human immunodeficiency virus-1 infection. Endocr Rev. 2010;31:79–97.
25. Munoz-Suano A, Hamilton AB, Betz AG. Gimme shelter: the immune system during pregnancy. Immunol Rev. 2011;241:20–38.
26. de Man YA, Dolhain RJ, van de Geijn FE, Willemsen SP, Hazes JM. Disease activity of rheumatoid arthritis during pregnancy: results from a nationwide prospective study. Arthritis Rheum. 2008;59:1241–8.
27. Østensen M, Villiger PM, Förger F. Interaction of pregnancy and autoimmune rheumatic disease. Autoimmun Rev. 2012;11:A437–46.
28. Latman NS. Relation of menstrual cycle phase to symptoms of rheumatoid arthritis. Am J Med. 1983;74:957–60.
29. de Man YA, Bakker-Jonges LE, Goorbergh CM, Tillemans SP, Hooijkaas H, Hazes JM, Dolhain RJ. Women with rheumatoid arthritis negative for anti-cyclic citrullinated peptide and rheumatoid factor are more likely to improve during pregnancy, whereas in autoantibody-positive women autoantibody levels are not influenced by pregnancy. Ann Rheum Dis. 2010;69:420–3.
30. Houtchens M. Multiple sclerosis and pregnancy. Clin Obstet Gynecol. 2013;56:342–9.
31. Vukusic S, Ionescu I, El-Etr M, Schumacher M, Baulieu EE, Cornu C, Confavreux C. The Prevention of Post-Partum Relapses with Progestin and Estradiol in Multiple Sclerosis (POPART'MUS) trial: rationale, objectives and state of advancement. J Neurol Sci. 2009;286:114–8.
32. Andreoli L, Fredi M, Nalli C, Reggia R, Lojacono A, Motta M, Tincani A. Pregnancy implications for systemic lupus erythematosus and the antiphospholipid syndrome. J Autoimmun. 2012;38:J197–208.
33. Chabbert-Buffet N, Amoura Z, Scarabin PY, Frances C, Lévy DP, Galicier L, Wechsler B, Blétry O, Piette JC, Gompel A. Pregnane progestin contraception in systemic lupus erythematosus: a longitudinal study of 187 patients. Contraception. 2011;83:229–37.
34. Ansar Ahmed S, Young PR, Penhale WJ. The effects of female sex steroids on the development of autoimmune thyroiditis in thymectomized and irradiated rats. Clin Exp Immunol. 1983;54:351–8.

35. Strieder TG, Prummel MF, Tijssen JG, Endert E, Wiersinga WM. Risk factors for and prevalence of thyroid disorders in a cross-sectional study among healthy female relatives of patients with autoimmune thyroid disease. Clin Endocrinol (Oxf). 2003;59:396–401.
36. Itsekson AM, Seidman DS, Zolti M, Alesker M, Carp HJ. Steroid hormone hypersensitivity: clinical presentation and management. Fertil Steril. 2011;95:2571–3.
37. M, AA, Ian C. Autoimmune progesterone dermatitis, Chapter 19 in Ed. Mark L et al. Treatment of Skin Disease : Comprehensive Therapeutic Strategies, Elsevier-Saunders, 2014.
38. Magen E, Feldman V. Autoimmune progesterone anaphylaxis in a 24 year old woman. Isr Med Assoc J. 2012;14:518–9.
39. Snyder JL, Krishnaswamy G. Autoimmune progesterone dermatitis and its manifestation as anaphylaxis: a case report and literature review. Ann Allergy Asthma Immunol. 2003;90:469–77. quiz 477, 571.
40. Baptist, A. P., and J. L. Baldwin, 2004, Autoimmune progesterone dermatitis in a patient with endometriosis: case report and review of the literature: Clin Mol Allergy, v. 2, p. 10.
41. Stranahan D, Rausch D, Deng A, Gaspari A. The role of intradermal skin testing and patch testing in the diagnosis of autoimmune progesterone dermatitis. Dermatitis. 2006;17:39–42.
42. Prieto-Garcia A, Sloane DE, Gargiulo AR, Feldweg AM, Castells M. Autoimmune progesterone dermatitis: clinical presentation and management with progesterone desensitization for successful in vitro fertilization. Fertil Steril. 2011;95:1121.e9–13.
43. Itsekson AM, Soriano D, Zolti M, Seidman DS, Carp HJ. Intradermal sex hormone desensitization for relief of premenstrual symptoms may improve the obstetric outcome of women with recurrent pregnancy loss. Gynecol Endocrinol. 2013;29:169–72.
44. Medeiros S, Rodrigues-Alves R, Costa M, Afonso A, Rodrigues A, Cardoso J. Autoimmune progesterone dermatitis: treatment with oophorectomy. Clin Exp Dermatol. 2010;35:e12–3.

Chapter 14
Progestogens in Non-gynecological Indications

Christa Nadjafi-Triebsch

1 Introduction

Progesterone is the most basic of all steroid hormones. All other steroid hormones are produced physiologically by modifying the progesterone molecule into glucocorticoids, mineralocorticoids, estrogens and androgens. Progesterone acts by agonizing the progesterone receptor. It is known that numerous body organs have estrogen or androgen receptors, explaining the differences between both sexes. Sex differences are seen in behavior patterns, cyclic responses of the hypothalamus, hair production and distribution, in addition to the differences in the sex organs. Glucocorticoid receptors are also found in virtually all organs of the body. Therefore, it should be no surprise that progesterone receptors are widely distributed and that progesterone affects numerous organs in both sexes as well as the more well known effects of progesterone on the uterus and reproduction.

The discovery that men are also dependent on progesterone was exciting and piqued my interest. When a lack of progesterone is found in men, many conditions can be improved with progesterone supplementation. Two examples are prostate hyperplasia and erectile dysfunction. Indeed Lee [1] has reported the successful treatment of prostate cancer with progesterone and testosterone. In addition to progesterone being produced by the Leydig cells, it is also produced in the central nervous system by the glial cells of the brain and spinal cord [2] and in the Schwann

C. Nadjafi-Triebsch, M.D.
Foederatio Medicorum Helveticorum (F.M.H.), P.O Box, CH-3000 Bern, 15, Switzerland

Aerztekammer Nordrhein/Germany (A.E.K.N.O.), Tersteegenstr.9, D-40474 Düsseldorf/Golzheim, Germany

German Society of Anti-Aging Medicine e.V. (G.S.A.A.M.), Vallstedter Weg 114, D-38268 Lengede, Germany

Swiss Society for Anti Aging Medicine and Prevention (S.S.A.A.M.P.), Industriestrasse 3, CH-6345, Neuheim/Zug

H.J.A. Carp (ed.), *Progestogens in Obstetrics and Gynecology*,
DOI 10.1007/978-3-319-14385-9_14

cells [3] of peripheral nerves. Therefore, progesterone supplementation has been used successfully in depression, sleep disorders, multiple sclerosis, spinal cord neurodegeneration, brain trauma, CVA and epileptic seizures.

Both a gene defect or an age-related decrease may lead to a lack of progesterone.

Adequate substitution of progesterone for both prophylaxis and the correction of several deficiencies in the male and also in children and for a number of neurological disorders is not only justified, but highly desirable.

Since the late 1980's a human identical form of progesterone has been available.

Micronization of progesterone has allowed the hormone to be administered orally rather than by deep intramuscular injection when dissolved in oil. As the objective of progesterone supplementation is to replace a lack of the physiological hormone, I only treat patients with bio-identical progesterone, and not synthetic progestogens.

However, as progesterone is not widely used outside of the accepted gynecological indications (described in other chapters in this book), appropriate control studies are urgently required.

I would like to dedicate this chapter to my patients, from whom I have learned what can be achieved when the lack of progesterone is corrected.

2 Progesterone in Males

Relatively little is known about normal values of hormones in men compared to the physiological levels in women. Although diminished levels of testosterone and DHEA are now often corrected with human identical hormones, the restoration of a healthy balance of hormones however, rarely achieves attention. As stated above the interaction of numerous hormones and enzymes is required in order to ensure an optimal hormone balance.

2.1 Biosynthesis of Sex Hormones

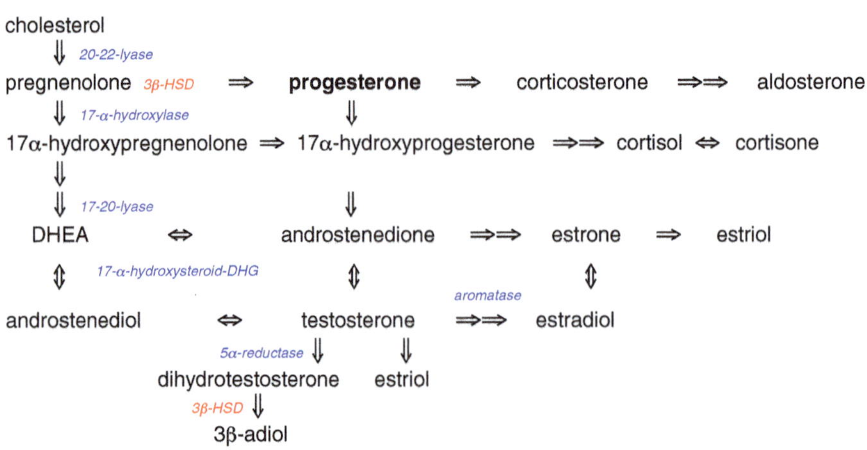

2.2 Progesterone Production in the Male

In men progesterone is produced in the Leydig cells of the testes, suprarenal gland, glial cells of the brain and the spinal cord [2] and in the Schwann cells [3] of peripheral nerves. Progesterone, Pregnenolone and their metabolites have been found in all of the above tissues. In the kidney, progesterone is metabolized to androgens.

2.3 Progesterone Receptors

The prostate is the male equivalent of the uterus and Skene's glands. Since both develop from the distal part of the Müllerian duct and the surrounding glands, it is not surprising that the prostate has progesterone receptors as well as estrogen and testosterone receptors [4]. There are two isoforms of progesterone receptors: PR-A and PR-B. The PR-B isoform is a full-length receptor, the PR-A isoform has 164 amino acids fewer than the PR-B receptor. The ratio of PR-A and PR-B levels in the target cells determine the type and extent of the progesterone effect [5]. PR-A is responsible for progesterone-dependent reproduction and the PR-B for normal differentiating effects, e.g. the breast [6]. PR-B is more active than PR-A and is cell-specific. PR-A suppresses transcription activities of other steroid hormone receptors, including ERα and PR-B. Progesterone has similar effects as 3β-Adiol, the agonist of ERβ, which should rather be renamed the 3β-Adiol receptor, since 3β-Adiol is not an estrogen. Progesterone receptors have been identified in the heart, liver, sperm, epithelial cells of the eyes, brain and nerves and in the prostate. There are also two isoforms of 3β-hydroxysteroid dehydrogenase. Type I in the placenta and skin and Type II in the suprarenal gland, ovaries and testes. The progesterone receptor mediated suppression of gonadotropin reduces LH, FSH and thus testosterone in the male. It reduces pulsatile frequency and has little affinity for the androgen receptor [7].

The non-genomic effects of progesterone are regulated in men by intracellular receptors in the membrane. High concentrations influence the membrane liquid directly. Membrane-dependent progesterone effects are cell capacitation and LH receptor expression, leading to testosterone synthesis in the Leydig cells and interactions with the GABA receptor complex for sedation and anesthesia. Interactions also occur in fatty tissue and the kidneys.

2.4 Biological Activities of Progesterone

The highest levels of progesterone are found in the saliva of newborn infants. During the first month of life progesterone falls to one third of its previous level, and after 2 years the circadian rhythm has become established with high morning peaks and low evening values [8]. Progesterone displays a wide spectrum of biological activities in multiple tissues.

These effects can be stimulating or restraining, depending on the respective tissue, dose, point in time of application and progesterone receptor distribution.

2.5 Progesterone Prevents Prostate Cancer

Hereditary prostate cancer is associated with a defect in the 3β-hydroxysteroid dehydrogenase gene [9], which codes for the enzyme needed to metabolize progesterone from pregnenolone leading to a progesterone deficiency. The same enzyme is needed to metabolize 3β-Adiol from dihydrotestosterone (DHT). Hence the DHT level remains in excessive concentrations and 3β-Adiol, an important protective hormone, is reduced.

Additionally the age-related decrease of progesterone leads to a gradual decrease of testosterone and an increase of cell growth promoting estradiol [1, 9]. When hormone levels diminish in the aging male, markers of inflammation increase. These markers include C-reactive protein (CRP), interleukin-1β, interleukin-6, TNFα and total and free PSA levels and the quotient of both. In patients with prostate hyperplasia receiving progesterone, these parameters of inflammation decrease steadily back to the normal range. Micturition problems improve or even disappear with progesterone supplementation as the prostate shrinks. However, the process takes approximately 9–12 months.

Progesterone as the natural 5α-reductase inhibitor controls the metabolism of testosterone to DHT. The deficiency of 3β-hydroxysteroid-dehydrogenase (whether hereditary or age-related) leads to a decrease of progesterone, thus leaving DHT unopposed. Since 3β-Adiol cannot be metabolized from DHT, an excessive amount of DHT remains to promote prostate hyperplasia [10]. Progesterone is also the natural aromatase inhibitor controlling the effect of growth-promoting estradiol. The luminal cells of prostatic epithelia show high amounts of ERβ, whereas ERα is found primarily in the basal cells. In prostate cancer ERβ is down-regulated and ERα is spread to the luminal cells. ERβ is reduced 10 times and PR-A and PR-B are deficient, whereas ERα remains unchanged, leaving 17β-estradiol functions unopposed [11]. Theses mechanisms may explain why progesterone deficiency triggers the development of prostate cancer.

2.6 Other Important Effects of Progesterone in the Male

Progesterone is needed for fertility in men, since progesterone increases the volume of the ejaculate and improves sperm motility. The addition of progesterone to the Percoll medium in cases of assisted reproduction significantly increases sperm motility [12].

Progesterone improves sexual performance in rats [13]. Consequently, progesterone should also improve sexual performance in men due to similarities to human neuroendocrine mechanisms. Indeed, patients receiving progesterone reported more frequent morning erections and distinctively improved sexual performance.

The anxiolytic effects of oral progesterone have been examined in a double-blind crossover study in 38 men. The Hamilton Anxiety Rating Scale was significantly reduced 4 h after progesterone administration and remained lower when examined after 9 and 24 h. Patients in the placebo arm of the trial only had a mild and non-significant reduction of the anxiety scale [14]. Additionally, the dose of antidepressants can be gradually reduced and often stopped completely [authors's series, unpublished].

The sleep-inducing effects of progesterone are mediated by allopregnanolone and pregnanolone, the metabolites of progesterone [15]. Most of my patients report that sleep is improved with progesterone supplementation. Part of the mechanism whereby progesterone improves sleep patterns may be due to progesterone increasing pulmonary gas exchange and reducing alveolar CO_2 pressure, leading to improved respiration and therefore to undisturbed sleep [16, 17]. Under progesterone therapy, many patients snore less or stop entirely. Whether sleep apnoea is also due to progesterone deficiency requires further investigation.

Progesterone supplementation reduces cholesterol and returns the lipoprotein levels to normal. Normalization of cholesterol levels takes place especially quickly if in addition to progesterone supplementation, hypothyroidism is also corrected with a thyroid preparation (I personally use Novothyral, Merck Serono, Switzerland). After correction of the hormone profile, there are patients who could stop statins completely.

It has been reported that patients with asthma can stop their medication if substituted with progesterone [18]. I have only treated one patient with asthma, but the other medication could be stopped. However, trials are needed in asthma.

The anti-inflammatory effects of progesterone have been reported in men with active arthritis [19]. Intra-articular injection to one knee in 12 men produced a local anti-inflammatory effect, which lasted for 3 months. An especially impressive positive effect was observed in two men with polyarthritis, who did not respond to other therapies. The authors suggested either an immunosuppressive effect on lymphocytes or induction of immunosuppressive glycoproteins or binding of the glucocorticoid receptor, to explain the anti-inflammatory effect. Ginanneschi et al. [20] have treated 16 women with carpal tunnel syndrome by local injection of 17α-hydroxyprogesterone caproate. Progesterone therapy was compared to corticosteroids. Corticosteroid therapy was followed by a 1 month pain-free period. However, the pain-free period was six months after injection of the long acting progesterone derivative.

Other important effects of progesterone are anesthesia, immune suppression, mild diuretic, anti-hypertensive [21], anti-convulsive [22], anti-oxidant actions [23] and bone formation [24]. Progesterone enhances thyroid function, normalizes blood sugar, zinc and copper levels.

2.7 Standard Values of Progesterone in Men

At present, no laboratory has published normal values of progesterone in men. Lee used 5–8 mg progesterone transdermally and considered a level of 400 pg/ml in saliva as optimal [1]. Unfortunately, at present, no laboratory in Switzerland uses saliva levels to determine hormone levels. Rimkus [25] has used a progesterone level of 4–10 ng/ml in serum, and supplements progesterone levels if under 4 ng/ml. If the serum value increases to 10 ng/ml, it can be considered normal, but in my experience a progesterone level of 12 ng/ml is better. None of my patients had a normal value of progesterone. The progesterone levels are age-dependent lower with increasing age. Using a dose of 100 mg oral micronized progesterone (Utrogestan, Bessins International, Belgium) at night from day 6 of the month until the end of each month, (with a break from day 1–5 of the month, to prevent down-regulation of the receptor) [1] serum values of progesterone increased to 3.5 ng/ml up to 12 ng/ml. With substitution of DHEA or testosterone, the progesterone level increased to 6 ng/ml.

3 Progesterone, the Neurotrophic Hormone

Progesterone is not only produced in the Leydig cells of the testes, the ovary and adrenal cortex, but also in the glial cells in the brain and spinal cord [2] and in the Schwann cells of the peripheral nerves [3], (with especially high levels in the sciatic nerve). Consequently progesterone has many important effects on the brain and nerves. Some effects are described below.

3.1 Multiple Sclerosis (MS)

MS is an autoimmune inflammatory disease affecting the central nervous system, with demyelinization and neurodegeneration. The condition is characterized by relapses and remissions. Progesterone increases the number of oligodendrocytes [26] in the brain. Oligodendrocytes produce the protein myelin. Optimal levels of progesterone or pregnenolone lead to a statistically significant increase of myelin sheaths [27]. Progesterone suppresses matrix metalloproteinases [28], which maintain the inflammatory plaques in multiple sclerosis. Consequently in pregnancy under high progesterone levels exacerbation of MS does not occur. Post partum relapses are considered to be induced by the decreased levels of these steroids [29]. Progesterone has been shown to reduce the inflammatory reactions commonly seen in MS. This is due to the direct effect of progesterone on astrocytes and microglia [30]. I have one female patient, aged 62 with MS, who has received progesterone supplementation (200 mg micronized progesterone from day 6 until the end of the month) for the last 2 years. She has had improvement of her symptoms and can now use stairs more easily. Trials are urgently needed.

3.2 *Brain Trauma and Stroke*

Progesterone has been compared to placebo in rats within 24 h after brain trauma as long ago as 2002 [31]. The rats received 4 mg/kg progesterone over 3–5 days. Edema of the brain and behavioral abnormalities were only prevented when progesterone was administered for at least 5 days. In 2007 Wright and co-workers [32] reported protection against necrotic damage and behavioral abnormalities caused by traumatic brain injury in 77 of 100 patients receiving progesterone within 11 h after brain trauma. In this series progesterone was administered intravenously (5 mg/kg/h/3days). The 30-day mortality rate was lower, when compared to 23 patients receiving placebo. In moderate traumatic brain injury a moderate to good recovery of brain functions under progesterone compared to placebo was achieved. Xiao et al. [33] treated 82 of 159 patients within 8 h after brain trauma with a Glasgow Coma Score less than or equal to 8 within 8 h. Progesterone was administered intramuscularly in a dose of 1 mg/kg body weight every 12 h for 5 days. Treated patients had a moderate to good recovery of brain functions compared to placebo patients. Progesterone treated patients had a statistically significant lower 6-months mortality compared to the 77 patients receiving placebo. Additionally, the average intracranial pressure 72 h and 7 days post trauma was lower in the progesterone treated group. No side effects were observed with progesterone!

In 2009 three important studies were published demonstrating that progesterone leads to less brain damage, improves the blood–brain barrier and reduces markers of inflammation in rats with stroke [34]. If vitamin D3 was also administered immediately after the stroke together with the progesterone, recovery was even better [35, 36]. These important publications led to a large multi-center trial in patients with brain trauma or stroke. The trial commenced in 2010, and is recruiting patients from 17 hospitals in 15 US states. The patients receive progesterone in a 96-h (**4 days**) continuous intravenous infusion. The trial is scheduled to terminate in 2015. The results are eagerly awaited.

In my opinion, patients with stroke or brain trauma could have a better prognosis if they would be treated already now with an intramuscular injection of 1 mg per kg body weight every 12 h for 5 days.

3.3 *Epilepsy*

As long ago as 1995, Herzog [37] published, that progesterone reduces complex partial seizures (CPS) and secondary generalized motor seizures (SGMS) in women with low serum mid-luteal progesterone levels (less than 5 ng/ml). Twenty-five women with CPS and SGMS received progesterone supplementation for 3 months. 18 (72 %) experienced a decline in the frequency of seizures for CPS ($p < 0.01$) and SGMS ($P < 0.02$) compared to the 3 months prior to therapy. Since 1995, there have been an additional 11 publications regarding the effect of progesterone in epilepsy [38–48].

3.4 Parkinson's Disease

The dopamine neurons of hemi-Parkinson rats displayed increased progesterone receptors A and B and a decrease in the estrogen receptor ERα, compared to initial pluri-potency during differentiation. Ninety-two per cent of the dopamine neurons had progesterone receptors [49]. It appears to be justified to use progesterone in Parkinson's disease, since human chorionic gonadotropin (hCG) increases testosterone levels with concomitant alleviation of the symptoms of Parkinson's disease [50]. Unfortunately progesterone was not assessed in that study.

3.5 Progesterone Protects the Spinal Cord from Neurodegeneration

Progesterone pellets were given to wobbler mice with pronounced neurodegeneration. Progesterone led to reduction of the cell vacuoles and the mitochondrial structure remained intact. In treated mice the grip strength was improved and survival was prolonged compared to the untreated mice [51].

3.6 Progesterone in ADD, ADHD and Asperger Syndrome

It has been reported, that when progesterone is given to progesterone-deficient pregnant women, the children are more advanced at 1 year and have greater academic achievement at 9–10 years and 17–20 years. The best academic results were found when the mothers received progesterone before the 16th gestational week, for longer than 8 weeks and over 5 g in total [52]. In 2001 Trotter [53] reported that progesterone and estradiol supplementation in premature babies to the levels they would have received in utero, led to normal psychomotor development, higher bone mineralization and reduction of lung diseases.

Platt [18] is of the opinion that progesterone deficiency is the cause for attention deficit disorder, and that progesterone deficiency together with insulin excess and noradrenalin excess leads to ADHD, as lack of progesterone, too much insulin and noradrenalin. Platt has reported on a 9-year-old boy with ADHD who lost his behavior disturbances when progesterone was administered. In 6 months he became one of the best pupils in class. I have treated a 12-year-old boy with low values of progesterone and testosterone (according to the Tanner classification [54]. He was also deficient in vitamin D and iron and had several food intolerances. He received 100 mg progesterone before bedtime from day 6 until the end of each month and daily iron and 1,000 IE vitamin D3. The serum levels of progesterone, estradiol, testosterone, DHEA-S and noradrenalin decreased into the normal range according to the Tanner classification [54].

After 3 months this youth received better grades at school. In the meantime I have treated ten more patients with similar good results. One patient with ADHD was previously treated with Ritalin®. After progesterone supplementation, Ritalin® could be stopped immediately. The lack of progesterone is due to a gene defect of 3β-hydroxysteroid dehydrogenase [1], the enzyme needed to produce progesterone from pregnenolone, hence the progesterone deficiency. In children with suspected ADD or ADHD, I assess the following parameters: FSH, LH, pregnenolone, progesterone, testosterone, dihydrotestosterone (DHT), DHEA-S, serotonin, melatonin, TSH, T3 and T4 (as there is often concomitant hypothyroidism [55]), noradrenalin (reduced in ADD and increased in ADHD), vitamin D (required for progesterone absorption), ferritin (as iron deficiency often co-exists), glucose and insulin.

Dietary modification with a drastic reduction of carbohydrates may also be helpful. These children receive 100 mg oral progesterone before bedtime from day 6 of the month until the end of each month, with a break from day 1 to 5. After 3 months a significant improvement in concentration and academic achievement is evident. These children not only have values of progesterone that are too low, but also very low values of testosterone. With progesterone therapy testosterone levels increase as manifested by penile growth. With these changes height increases, as does self-esteem. Iron supplementation can usually be stopped after 3–5 months, as iron can be absorbed more easily when the hormone profile is corrected. In most ADD and ADHD children hypothyroidism is present due to a gene defect. Therefore thyroid hormones supplementation is required together with progesterone on a lifelong basis.

Hypothyroidism is often the cause for acne in ADD and ADHD children [55] and acne abates after approximately 4 months of treatment with thyroid hormones and progesterone.

Dr. Felix Horn [personal communication] has treated two children with Asperger syndrome (an autism spectrum disorder) using the same tests for ADD and ADHD as listed above. Progesterone values were found to be extremely low. With progesterone supplementation both patients improved considerably with regard to communication with other children.

References

1. Lee JR. Hormone balance for men. What your doctor may not tell you about prostate health and natural hormone supplementation. One to One Inc. P.O. Box 84900 Phoenix, AZ 85071;2007, (800) 528–0559.
2. Inue T, Akahira JL, Suzuki T, Darnel AD, Kaneko C, Takaha K, et al. Progesterone production and actions in the human central nervous system and neurogenic tumors. J Clin Endocrinol Metab. 2002;87(11):5325–31.
3. Koenig HL, Schuhmacher M, Ferzaz B, Do Thi AN, Ressouches A, Guennoun R, Jung-Testas I, Robel P, Akwa Y, Baulieu EE. Progesterone synthesis and myelin formation by Schwann cells. Science. 1995;268(5216):1500–3.
4. Srinivasan G, Cambell E, Beshirelahi N. Androgens, estrogens and progesterone receptors in normal and aging prostates. Microsc Res Tech. 1995;30(4):293–304.

5. Inue T, Akahira JL, Takeyama J, Suzuki T, Darnel AD, Kaneko C, Kurokawa Y, Satomi S, Sanano H. Spacial and topological distribution of progesterone receptor A and B isoforms during human development. Mol Cell Endocrinol. 2001;182(1):83–9.
6. Ariga N, Suzuki T, Moriya T, Kimura M, Inoue T. Progesterone receptor A and B isoforms in the human breast and its disorders. J Endocrinol Metab. 2004;89(3):1429–42.
7. Brady BM, Anderson RA, Kinniburgh D, Baird DT. Demonstration of progesterone receptor-mediated gonadotrophin suppression in the human male. Clin Endocrinol (Oxf). 2003; 58(4):506–12.
8. Gröschl M, Rauh M, Dörr H-J. Circadian rhythm of salivary cortisol, 17α- hydroxy progesterone and progesterone in healthy children. Clin Chem. 2003;49:1688–91.
9. Guerini V, Sau D, Scaccianoce E, Rusmini P, Ciana P, Maggi A, Martini PGV, Katzenellenbogen BS, Martini L, Motta M, Poletti A. The androgen derivative 5α-androstane-3β, 17β-diol inhibits prostate cancer cell migration through activation of the estrogen receptor β subtype. Cancer Res. 2005;65(12).
10. Mauvais-Jarvis P, Kuttenn F, Baudet N. Inhibition of testosterone conversion to dihydrotestosterone in men treated percutaneously by progesterone. J Clin Endocrinol Metab. 1974; 38(1):142–7.
11. Bonkhoff H, Fixemer T. Bedeutung der Östrogene und ihrer Rezeptoren für die Entstehung und Progression des Prostatakarzinoms. Der Pathologe. 2005;26(6):461. 46.
12. Andersen ML, Tufic S. Does male sexual behavior require progesterone? Brain Res Rev. 2006;51(1):36–43.
13. Witt DM, Young LJJ, Crews D. Progesterone modulation of androgen dependent sexual behavior in male rats. Physiol Behav. 1995;57(2):307–13.
14. De Lignieres B, Beausang A, Faltot M. Acute anxiolytic effect of oral administration of progesterone in men: a double-blind crossover study. Second international symposium of premenstrual, postpartum and menopausal mood disorders. September Kiahaj Island, SC, USA; 1987.
15. Fries E, Tagaya H, Trachsel L, Holsboer F, Rupprecht R. Progesterone- induced changes in sleep in male subjects. Am J Physiol. 1997;272:E885–91.
16. Saaressranta T, Polo-Kantola P, Irjala K, Helenius H, Polo O. Respiratory insuffiency in postmenopausal women: Sustained improvement of gas exchange with short-term medroxyprogesterone acetate. Chest. 1999;115(6):1581–87.
17. Saaressranta T, Polo-Kantola P, Rauhala E, Polo O. Medroxyprogesterone in postmenopausal females with partial upper airway obstruction during sleep. Eur Respir J. 2001;18:989–95.
18. Platt ME. The miracle of bio-identical hormones. Clancy Lane Publishing, Rancho Mirage, USA. ISBN 978-0-9776683-2-8.
19. Cuchacovich M, Tchernitchin A, Gatica H, Wurgaft R, Valenzula C, Cornejo E. Intraarticular progesterone: effects of a local treatment for rheumatoid arthritis. J Rheumatol. 1988; 15(4):561–65.
20. Ginanneschi F, Milani P, Filippou G, Mondelli M, Frediani B, Melcangi RC, Rossi A. Evidences for antinociceptive effect of 17-alpha-hydroxy progesterone caproate in carpal tunnel syndrome. J Mol Neurosci. 2012;47(1):59–66.
21. Rylance PB, Brincat M, Lafferty K, De Trafford JC, Brincat S, Parsons V, Studd JW. Natural progesterone and antihypertensive action. BMJ. 1985;290:13–4.
22. Reddy DS. Role of neurosteroids in catamenial epilepsy. Neurotherapeutics. 2009;6(2): 392–401.
23. Tranquilli AL, Mazzini L, Cugini AM, Cester N, Garzetti GG, Romanini C. Transdermal estradiol and medroxyprogesterone acetate in hormone replacement therapy are both antioxidants. Gynecol Endocrinol. 1995;9(2):137–41.
24. Burnett CC, Reddi AH. Influence of estrogen and progesterone on matrix-induced endochondral bone formation. Calcif Tissue Int. 1983;35:609–14.
25. Rimkus V (2000) Der Mann im Wechsel seiner Jahre. Arche Noah, Musik- und Buchverlag D-86971 Peiting. 3. Auflage ISBN 10:3-86733-000-X.

26. Baulieu EE, Schuhmacher M. Neuro steroids, with special references to the effect of progesterone on myelination in peripheral nerves. Mult Scler. 1997;3:105–12.
27. Schuhmacher M, Guennoun R, Robert F, Carelli C, Gago N, Ghoumari A, Gonzales Deniselle MC, Gonzales SL, Ibanez C, Labombarda F, Coirini H, Beaulieu EE, De Nicola AF. Local synthesis and dual actions of progesterone in the nervous system: Neuroprotection and myelination. Growth Horm IGF Res. 2004;14(Suppl A):S18–33.
28. Chen JL, Lin QH, Fang XL, Tao GS, Huang FY. Effect of progesterone on the secretion of matrix metalloproteinase-2 and matrix metalloproteinase-9 in human ectopic endometrial stromal cells. Zhong Nan Da Xue Bao Yi Xue Ban. 2005;30(3):307–11.
29. Confavreux C, et al. Rate of pregnancy-related relapse in multiple sclerosis. N Engl J Med. 1998;339:285–91.
30. Garay LI, et al. Progesterone down-regulates spinal cord inflammatory mediators and increases myelination in experimental autoimmune encephalomyelitis. Neuroscience. 2012;226:40–50.
31. Shear DA, Galani R, Hoffman SW, Stein DG. Progesterone protects against necrotic damage and behavioral abnormalities caused by traumatic brain injury. Exp Neurol. 2002; 178(1):59–67.
32. Wright DW, Kellermann AL, Hertzberg VS, et al. Pro TECT: a randomized clinical trial of progesterone for acute traumatic brain injury. Ann Emerg Med. 2007;49(4):391–402.
33. Xiao G, Wei J, Yan W, Lu Z. Improved outcomes from administration of progesterone for patients with acute severe traumatic brain injury: a randomized controlled trial. Crit Care. 2008;12(2):R61.
34. Ishrat T, Sayeed I, Atif F, Stein DG. Effects of progesterone administration on infarct volume and functional deficits following permanent focal cerebral ischemia in rats. Brain Res. 2009;1257:94–101.
35. Cekic M, Cutler SM, Valandingham JW, Stein DG. Vitamin D deficiency reduces the benefits of progesterone treatment after brain injury in aged rats. Neurobiol Aging. 2011; 32(5):864–74.
36. Wang J, Jiang C, Li X, et al. The protective mechanism of progesterone on blood–brain barrier in cerebral ischemia in rats. Brain Res Bull. 2009;9(6):426–30.
37. Herzog AG. Progesterone therapy in women with complex partial and secondary generalized motor seizures. Neurology. 1995;45:1660–2.
38. Bauer J, Hocke A, Elger CE. Katameniale Anfälle: Eine Analyse. Nervenarzt. 1995;66: 760–9.
39. Herzog AG. Three patterns of catamenial seizures. Epilepsia. 1997;38:1082–8.
40. Bauer J, Burr W, Elger CE. Seizure occurrence during ovulatory and anovulatory cycles in patients with temporal lobe epilepsy: a prospective study. Eur J Neurol. 1998;5:83–8.
41. Herzog AG. Progesterone therapy in women with epilepsy: a 3-year follow-up. Neurology. 1999;52(9):1917–8.
42. Edwards HE, Epps T, Carlen P, Mac Lusky N J. Progesterone receptors mediate progesterone suppression of epilepti form activity in tetanized hippocampal slices in vitro. Neuroscience. 2000;101(4):895–909.
43. Herzog AG, Friedman MN, Freund S, Pascal-Leone A. Transcranial magnetic stimulation evidence of a potential role for progesterone in the modulation of premenopausal corticocortical inhibition in a woman with catamenial seizure exacerbation. Epilepsy Behav. 2000; 2(4):367–9.
44. Herzog AG, Frye CA. Seizure exacerbation associated with inhibition of progesterone metabolism. Ann Neurol. 2003;53(3):390–1.
45. Bauer J (2004) Warum treten epileptische Anfälle kurz vor der Periode auf? Informationspool Epilepsie. Dt. Gesellschaft für Epileptologie.
46. Reddy DS. Role of neurosteroids in catamenial epilepsy. Epilepsy Res. 2004;62(2–3): 99–118.
47. Harden CL, Herzog AG, Nokolov BG, Koppel BS, Christos PJ, Fowler K, Labar DR, Hauser WA. Hormone replacement therapy in women with epilepsy: a randomized, double-blind, placebo-controlled study. Epilepsy. 2006;47(9):1447–51.

48. Reddy DS, Rogawski MA. Neurosteroid replacement therapy for catamenial epilepsy. Neurotherapeutics. 2009;6(2):392–401.
49. Diaz NF, Gueerra-Arraiza CH, Diaz NE, Salazar P, Molina-Hernandez A, Camacho-arroyo I, Velasco I. Changes in the content of estrogen α and progesterone receptors during differentiation of mouse embryonic stem cells to dopamine neurons. Brain Res Bull. 2007;73:75–80.
50. Gomula A. Successful treatment of parkinsonism with HCG: preliminary findings. Presented at the 4th ISSAM conference in Prague. 2004.
51. Gonzales Demiselle MC, Lopez Costa JJ, Gonzales SL, Labombarda F, Garay L, Guennounoun R, Schuhmacher M, De Nicola AF. Basis of Protection in spinal cord neurodegeneration. J Steroid Biochem Mol Biol. 2002;83(1–5):199–209.
52. Trotter A, Maier L, Pohlandt F. Management of the extremely preterm infant: is the replacement of estradiol and progesterone beneficial? Pediatr Drugs. 2001;3(9):629–37.
53. Trotter A, Bokelmann B, Sorgo W, Bechinger-Kornhuber D, Heinemann H, Schmücker G, Oesterle M, Köhntop B, Brisch KH, Pollandt F. Follow-up examination at the age of 15 months of extremely preterm infants after postnatal estradiol and progesterone replacement. J Clin Endocrinol Metab. 2001;86(2):601–3.
54. Tanner M. Variations in the pattern of pubertal changes in boys. Arch Dis Child. 1970;45(239):S13–23.
55. Stark M. Hypothyroidism Type 2: the epidemic. Revisedth ed. Columbia, MO: Mark Starr Trust; 2013.

ERRATUM TO

Progestogens in Non-gynecological Indications

Christa Nadjafi-Triebsch

H.J.A. Carp (ed.), *Progestogens in Obstetrics and Gynecology*,
DOI 10.1007/978-3-319-14385-9_14

DOI 10.1007/978-3-319-14385-9_15

The last sentence on page 198 "The serum levels of progesterone, estradiol, testosterone, DHEA-S and noradrenalin decreased into the normal range according to the Tanner classification [54]." should be replaced with following sentence:

The serum levels of progesterone, estradiol, testosterone, DHEA-S and noradrenalin increased in patients with ADD into the normal range according to the Tanner classification [52]. In patients with ADHD however, noradrenalin is decreased into the normal range, whereas progesterone, estradiol, testosterone and DHEA-S is, as in the ADD-patients, increased into the normal range.

The forgotten source for the graph on page 192, should read as follows:

By John R. Lee, M.D. with Virginia Hopkins: What your doctor may not tell you about menopause: the breakthrough book on natural progesterone. ISBN 0-446-67144-4, 1996.

The online version of the original chapter can be found at
http://dx.doi.org/10.1007/978-3-319-14385-9_14

H.J.A. Carp (ed.), *Progestogens in Obstetrics and Gynecology*,
DOI 10.1007/978-3-319-14385-9_15

FSC
www.fsc.org
MIX
Papier | Fördert gute Waldnutzung
FSC® C083411

Kolibri 360 GmbH
Ferdinand-Jühlke-Straße 7
99095 Erfurt, Deutschland
produktsicherheit@kolibri360.de